W0071658

The Staphylinidae (rove beetles) of Britain and Ireland

Part 5: Scaphidiinae, Piestinae, Oxytelinae

Derek A. Lott

Stenus Research

Colour plates prepared by

James Turner

Amgueddfa Cymru –
National Museum Wales

DEREK A. LOTT

© Royal Entomological Society 2009.

Published for the Royal Entomological Society
The Mansion House
Bonehill
Chiswell Green Lane
Chiswell Green
St Albans
AL2 3NS
www.royensoc.co.uk

By the   Field Studies Council
The Annexe
Preston Montford Lane
Shrewsbury
SY4 1DU
www.field-studies-council.org

ISBN: 978 0 901546 90 6

All rights reserved. No part of this book may be
reproduced or translated in any form or by any means,
electronically, mechanically, by photocopying or
otherwise, without written permission from the
copyright holders.

Contents

Abstract

Illustrated keys are provided for the identification of the subfamilies of British Staphylinidae. Three subfamilies Scaphidiinae, Piestinae and Oxytelinae, which form the Oxyteline group, are covered in full in this volume. A checklist is provided for the species covered in this volume. Keys cover the adults of all species that have been reliably recorded as established breeding populations in the British Isles.

Keys to genera and species are given along with species accounts. The species accounts provide key characters, information on differences between similar species, summaries of habitats, geographical distribution, and biology, where such information is available.

Acknowledgements

For loans of specimens and access to collections in their care I thank Roger Booth (Natural History Museum, London), Anona Finch (Leicestershire Museums Service), Mike Wilson (National Museum of Wales) and Tony Drane. For feedback on test keys and comments on various drafts of the manuscript I am grateful to Tony Allen, Mark Telfer, Roger Booth, Colin Welch, Bob Marsh, Martin Collier, Andrew Duff and Steve Bolchover. Peter Hammond, Gyorgy Makranczy, Roger Booth, Mark Telfer and Colin Johnson have at various times given me tips on identification characters that are incorporated into the identification keys. More specifically, Mark Telfer provided the characters used in the couplet separating *Carpelimus incongruus* Steel and *C. zealandicus* (Sharp). Andrew Duff provided the template for typing the manuscript, when he initiated the *Beetles of the British Isles* project and later became Coleoptera representative for the RES Handbooks Steering Group. I am most grateful to Tony Allen for proof-reading the manuscript. I thank Rebecca Farley (Field Studies Council) for her help in preparing the manuscript and Mike Wilson (National Museum of Wales) for his editorial guidance.

Introduction

With over 55,000 species described worldwide and over 1,000 of these recorded in Britain and Ireland, the Staphylinidae represent one of the most species-rich families in the animal kingdom. They constitute around 5% of all insect species and around one quarter of the beetle species recorded in the British Isles. Consequently, they are one of the most commonly encountered invertebrate groups in habitats ranging from the intertidal zone to mountain summits. As such, they should not be ignored in any investigation of biodiversity at whatever scale, site-based, regional or national.

Their importance to biodiversity in terms of sheer numbers of species is supplemented by their interest to students in a variety of different disciplines. Amateur naturalists will find them a rewarding group to study, because they have plenty of rare species with cryptic habitats to test their ingenuity. Ecologists will find that they have a wide range of habitats and that they include species with specialised life histories and obligate dependencies on other taxa. Conservationists will find that many species are restricted in their distribution by particular environmental factors. Coupled with their high species richness, this makes Staphylinidae a particularly useful group for monitoring community responses to environmental change. Taxonomists will find that species limits in some genera still need to be worked out, even in this country, while the huge radiations of species in several subfamilies provide fertile ground for exploring evolutionary trends. Their chemical defence systems have attracted the attention of biochemists (see *e.g.* Dettner, 1993), while their various adaptations to their environment have attracted attention from both functional morphologists (see *e.g.* Betz *et al.*, 2003) and behavioural ecologists (see *e.g.* Betz, 1999).

Despite their evident interest, the study of Staphylinidae has lagged behind many other insect groups, because they have the reputation of being difficult to identify. The large number of species and the taxonomic and nomenclatural confusion affecting several genera have deterred many students from following up their initial enthusiasm for the group. In fact, Staphylinidae are relatively easy to identify to species compared to other beetles. The male and sometimes female genitalia tend to be highly distinctive and easy to examine and their exposed abdominal tergites often carry additional specific characters that are hidden in other families. Recent work on the taxonomy of central European Staphylinidae (see Assing & Schülke, 2007) has cleared up at least some of the confusion that used to confound attempts at their identification in Britain and Ireland. This book is the first of a series of identification keys that aims to address the neglect of Staphylinidae by incorporating recent advances in the taxonomy of Staphylinidae into simple, easy-to-use keys.

The keys cover the adults of all species that have been reliably recorded as established breeding populations in the British Isles. The British Isles are here held to consist of three island groups: Britain, Ireland and the Isle of Man. The island of Britain consists of three countries: England, Wales and Scotland. Scotland is particularly rich in offshore islands including the Hebrides, the Orkney Isles and the Shetland Isles. Biogeographically, the Channel Islands are not part of the British Isles and some beetles that occur there may not be included in the keys.

Morphology

Most adult Staphylinidae can be easily recognised by the short wing cases (elytra) leaving five to six well sclerotised abdominal tergites exposed. The Scaphidiinae and Scydmaeninae are exceptional among British Staphylinidae in having elytra that cover most of the abdomen, although they do retain well sclerotised terminal tergites. The complex wing-folding patterns used to accommodate the wings under the relatively short elytra are another distinguishing feature of the family that they share only with the Silphidae within the Coleoptera. Except for the subfamilies Micropeplinae and Pselaphinae, the abdomen is relatively flexible compared with species in other beetle families with exposed abdominal tergites.

In British species the body length ranges from below 1 mm to 35 mm. Body length is difficult to measure precisely, because the abdominal segments are telescopic and the whole abdomen expands in alcohol or shrinks when dried. Consequently, some latitude should be allowed when comparing measurements with values given in the keys.

Surface colour can be derived from pigments or physical characters of the surface that produce metallic sheens. Pigments can deteriorate over time, especially if exposed to light, and this leads colours to fade, while prolonged storage with killing agents can cause colours to darken. All these factors should be kept in mind when examining old specimens. Teneral adults are individuals that have recently emerged from pupation. Their pigments are not properly developed and they look paler than mature adults. They can be recognised in the Staphylinidae by their soft cuticles, which buckle when pressed lightly with a needle.

Fig. 1 shows the morphological terms most commonly used in the identification keys. Wherever possible, the keys concentrate on characters visible on the upper (dorsal) surface or at least visible from the side (lateral perspective). The term disc is used in the keys to indicate the central portion of body parts such as the head, pronotum and each elytron. The head, pronotum, elytra and abdominal tergites all have features that are important for identification and where these are mentioned in the keys, they should be examined on the disc, unless otherwise indicated. Microsculpture refers to the microscopic lines on the surface of the chitin that resembles fingerprints on the human skin. Puncturation refers to punctures or minute pits on the surface that normally contain sensory hairs. Pubescence refers to the general arrangement of hairs, whose length and orientation can provide useful diagnostic characters. Longer bristles (setae) may be present in addition to this pubescence. Important features of the head for identification include the length of the temple relative to the diameter of the eye. The mouthparts are produced forwards and the maxillary palps should be set so that they are easily visible, because they often provide useful characters for identification. Simple eyes (ocelli) are visible in Omaliinae and some Proteininae. The antennae are almost always 11-segmented, rarely 10- or 9-segmented with one genus (Pselaphinae: *Claviger*) 6-segmented. The relative lengths of individual antennal segments can be useful key characters. Antennal segments also vary in general shape from transverse through quadrate to elongate, where transverse means wider than long and elongate means longer than wide. However, great care is needed in determining the shape of a segment, whose appearance can often be affected by slight variations in the way that the antennae are set. In addition the shapes of antennal segments often display some infraspecific variation.

Important characters for identification on the pronotum, scutellum and elytra include general shape, impressed lines and furrows. The elytra meet along a suture and the adjacent impressed line is called the sutural stria. Lines of punctures on the elytra are also termed striae. Wing length

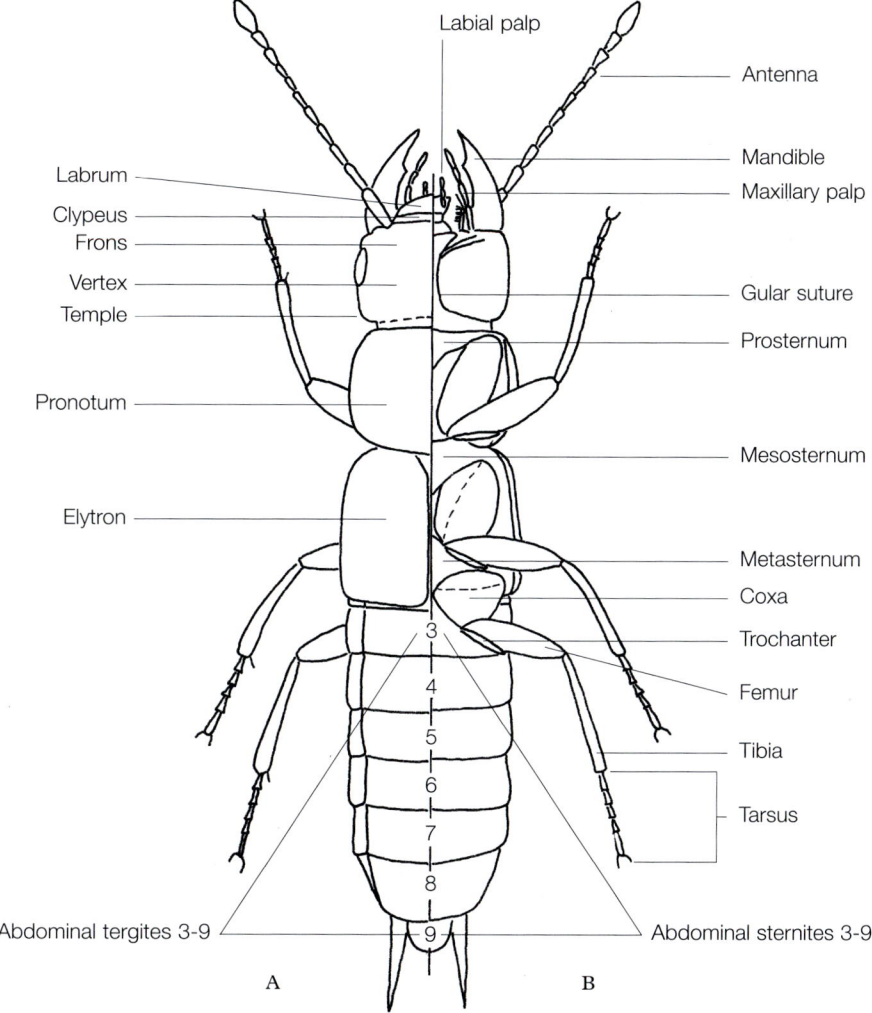

Figure 1. Staphylinid beetle (schematic) showing principle structure (A, upperside; B, underside).

can vary considerably both within and between species, but has not been used very much in the keys, so specimens can be prepared with their wings tucked up beneath the elytra. Indeed, if they are left covering the abdomen, they could obscure useful identification characters.

The tibiae can carry spines and bristles and the tarsi can vary in the number of visible segments. However, this last character can be difficult to appreciate, unless the tarsi are carefully prepared.

As a corollary of the evolution of short elytra, the Staphylinidae have developed chitinous abdominal tergites and these frequently provide important characters for identification. Unfortunately two different numbering systems have been applied to abdominal tergites and sternites and this can lead to confusion. As in all Coleoptera, the first true abdominal segment has disappeared, while the tergite of the second is normally hidden under the elytra. Consequently, the first visible tergite of most species is that of the third segment, referred to in the keys as tergite III. Many popular keys refer to tergites III to VIII as the 1st to 6th visible tergites. This device largely avoids the threat of confusion to students who may not be aware that the 1st visible tergite is technically tergite III. However, tergite III is not always the first visible tergite. In the Pselaphinae, tergite IV is the first visible tergite, so that there is one less visible tergite than in most other Staphylinidae, while in some genera of Oxytelinae, tergite II is relatively well developed and visible, although it lacks some of the features of the succeeding segments. In most species, tergites III to VI are very similar. Tergite VII is often slightly different and the corresponding sternite VII sometimes carries secondary sexual characters. Segment VIII is markedly different from preceding segments. It lacks a raised margin and both the tergite and sternite are much narrower and generally taper toward the apex. In some subfamilies sternite VIII can be used to sex specimens. In these subfamilies the apical margin is either incised or concave in the male, but simply rounded in the female. Segments IX and X are generally referred to as the genital segments. Their sclerites are highly modified and may or may not be visible at the tip of the abdomen, often depending on the preparation of the specimen.

Examination of the male genitalia (aedeagi) can be critical for separating closely related species. In the primitive form the aedeagus consists of a median lobe attached to two symmetrical parameres that have various degrees of free movement. In some genera the parameres are fused into a single piece or even missing altogether. The median lobe can contain internal sclerites, normally visible from the outside, and these can provide important characters for species identification. In these keys, the ventral aspect of the aedeagus refers to the normally concave surface to which the parameres are attached. Some works use the opposite convention. In females the spermatheca is well sclerotised in some groups, but is only used as a key character for the Aleocharinae.

Biology

The common English name for Staphylinidae is rove beetles, a term reportedly in use in the 18th century (Barbut, 1781). It is possible that this name is linked to the Norse word *rov* meaning prey, loot or booty rather than any wandering habit. In Danish they are called *rovbiller*, literally beetles of prey. The majority of Staphylinidae are commonly regarded as predators, but some are known to feed on fungi (Lipkow & Betz, 2005), algae (Bro Larsen, 1936), pollen (Newton & Thayer, 1995), dead plant material (Gildenkov, 2001a) and dung (Hinton, 1944). Even species commonly supposed to be predators may feed largely as scavengers (Good & Giller, 1991) or be more omnivorous than is often presumed. Undoubted predators as adults include large-eyed species of *Stenus* (Steninae) and *Quedius* (Staphylininae). *Quedius* species are among the few Staphylinidae that will kill and eat other Staphylinidae and this should be borne in mind when tubing live material in the field.

Most temperate species are believed to go through one generation per year, although there is some evidence that a few species can breed more than once. Kasule (1968) recognised several categories of life history among British Staphylinidae including species with summer larvae and species with winter larvae, but also found that some species had a more extended breeding season. It is evident that individual species can be quite plastic with regard to their life history. Whatever their breeding season, many species can be found as adults throughout the year. Some species are winter-active. In general, the habits of the larvae are poorly understood. The larvae of several species of *Aleochara* (Aleocharinae) are known to be specialist parasitoids of Diptera.

Staphylinidae tend to have similar habitats to Carabidae, but their adult body form represents a completely different solution to moving through their environment. Many adult Carabidae are wedge-pushers that move through the soil and leaf litter using brute force, whereas the elongate, flexible bodies of Staphylinidae allows them to weave through soil crevices, litter and tangled vegetation. Riparian species in the genera *Stenus* (Steninae) and *Ischnopoda* (Aleocharinae) that move over the flat, uncluttered surface of bare sand tend to have long legs and tarsi for fast running. Conversely, much shorter legs are found on species that habitually live in litter and well vegetated ground.

The ecology of coastal Staphylinidae in Britain has been reviewed by Hammond (2000), while Lott (2003) dealt with wetland and riparian Staphylinidae. Information on the autecology of saproxylic Staphylinidae has been compiled by Alexander (2002).

Collection methods and preservation

Staphylinidae can be collected using similar methods to those that are effective for Carabidae. Pitfall trapping, looking under stones, dissecting grass tussocks and sieving litter are invariably successful in finding large numbers. Many serious students of Staphylinidae use sieves sewn into canvas bags that can process large quantities of litter. Other productive techniques include Winkler extractors, interception traps and car nets, but light trapping usually turns up only the occasional specimen of interest in Britain and Ireland. Species with more restricted habitats require more specialised techniques and the reward of finding rarely recorded species using novel methods is part of what makes the study of Staphylinidae such fun. Species with specialist habitats can be found under bark and in rotten wood, fungal fruiting bodies, shingle banks, intertidal rock crevices, dung, carrion, underground mammal nests, ant nests and hornet nests.

Killing specimens with ethyl acetate keeps them relaxed and dry so that they can be easily examined and dissected, although this treatment largely prohibits subsequent DNA analysis. Material stored in alcohol or dilute acetic acid should be dried out in order to properly examine pubescence and surface microsculpture. Specimens stored in 70% ethanol sometimes become too brittle for easy examination or dissection and 50% ethanol is often more effective for short-term storage. However, for long-term storage there is a danger that specimens will fall apart at this dilution. For DNA analysis specimens should be killed in 95% ethanol and stored at low temperatures prior to processing. It is common practice to store rove beetles dry, by mounting them on good-quality card. Careful specimen preparation is critical in making a correct identification and this is best done when the specimen is relaxed. Above all, care should be taken to preserve the pubescence in good condition. Economical use of ethyl acetate for killing will avoid the accumulation of patches of grease on the surface that damage pubescence and obscure microsculpture and puncturation. Minimal use of glue for carding is also advisable for similar reasons.

Dissection of the abdomen to extract the genitalia is relatively easy in fresh, relaxed specimens before they have been carded. Museum specimens that have dried out can be relaxed by soaking them for a few minutes in water just below the boiling point. This is usually sufficient to permit easy dissection. In British species, except for Pselaphinae, it is a fairly simple process to insert a hook-tipped needle between tergite VII and VIII while holding down the rest of the specimen with a second mounted needle and pull off the last segment into a drop of water on a microscope slide. The genitalia can then be teased free. In some larger species, it may be necessary to cut through connective tissue to release the genitalia. Mounted needles can easily be customised using headless entomological pins and small corks or needle holders. Structures that are weakly sclerotised or that contain diagnostic internal sclerites should be stored in a liquid medium such as dimethyl hidantoin formaldehyde (DMHF) or Euparal. Many entomologists store genitalia on an acetate slip mounted just below the specimen card for easy examination under a microscope using transmitted light. However, the shelf life of acetate is limited and this is probably a risky stratagem for long-term storage. Mounting the genitalia on the same card as the specimen is acceptable for examination using incident light. The genitalia should always be mounted in a standard orientation suitable for displaying the diagnostic characters for a particular species. It is advisable to adopt a standard protocol for arranging dissected sclerites and structures on the card for easy comparison of specimens in future examinations.

Further information on both collecting methods and preparation techniques can be found in Cooter & Barclay (2006).

Identification

A binocular microscope is needed to appreciate many of the characters used in the keys. Most of them can be seen using 10x to 40x magnification, but it may be useful to go up to 60x or even higher to adequately examine surface microsculpture in smaller species. The quality of the light source is more important than the optical quality of the lens. Characters such as surface microsculpture and puncturation that are frequently used in the keys are best appreciated using a diffuse light. Daylight fluorescent tubes and frosted glass bulbs mounted in ordinary desktop lamps are suitable. Expensive fibre-optic light sources are virtually useless, because they create highlights that obscure surface microsculpture. For best results, the light should be directed at an angle to the specimen and the specimen moved round on a block of cork or plastazote until the required character receives the best illumination. This is particularly important for examining pubescence, because the hairs become invisible when the light is at the wrong angle. In this book illustrations of aedeagi and their internal sclerites were drawn using incident light. Extra internal structures can be seen using transmitted light from sub-stage illumination, but this can make the aedeagus look quite different.

The keys should be used to arrive at a provisional identification. Identifications of unfamiliar species should ideally be confirmed before any publication, especially if they are rare. One of the best ways to confirm an identification is to compare the specimen with a well-maintained reference collection in a museum.

Notes on keys

Each couplet in the keys presents a choice that leads either to a specific identification or to the next couplet. Characters given in square brackets are not exclusive to that side of the couplet. Overlap of characters between each side of the couplet is indicated in the text by the use of terms such as "on average", "generally" and "usually". The keys to large genera are often accompanied by diagrams that compare important features, such as genitalia, across several species. Further information on differences between similar species is given in the species accounts together with summaries of habitats, geographical distribution, and biology, where such information is available. Flight activity is indicated by reference to recorded captures in light traps or flight interception traps etc. For most subfamilies, information on species distributions within the British Isles is based on information kindly provided from the national recording scheme by Peter Hammond. Notes on the distribution of higher taxa and the number of species they contain are largely taken from Herman (2001).

Glossary

Aedeagus (-i): male genitalia

Apical: toward the end / extremity

Basal: toward the point of articulation

Chitin: material forming the exoskeleton

Clypeus: area of head immediately above the labrum

Coxa (-ae): part of leg (see Fig. 1)

Disc: the central portion of a body part such as the head, pronotum and each elytron

Dorsal: upper

Elongate: longer than wide

Elytron (-a): wing case (see Fig. 1)

Femur (-ora): part of leg (see Fig. 1)

Frons: area of head between and immediately in front of the compound eyes

Genital segments: internal abdominal segments IX and X

Labrum: mouthpart (see Fig. 1)

Lateral: side

Median lobe: central portion of aedeagus

Microsculpture: microscopic lines on the surface of the chitin

Ocellus (-i): simple eyes on vertex

Palps: mouthpart appendages (see Fig. 1)

Paramere: appendage of aedeagus

Pronotum: dorsal plate of prothorax (see Fig. 1)

Pubescence: covering of hairs

Puncturation: punctures on the surface that normally contain sensory hairs

Quadrate: as long as wide

Rugose: with uneven surface due to confluence of punctures

Sclerite: any specific plate or structure made of chitin

Scutellum: area of mesothorax often visible where elytra meet (see Fig. 1)

Setae: longer individual hairs or bristles

Spermatheca: sperm storage organ in female

Sternite: ventral plate (used mainly in keys for abdominal segments)

Stria (-ae): Line of punctures or impressed line

Sutural stria: impressed line adjacent to elytral suture

Tarsus (-i): foot (see Fig. 1)

Temple: side of head behind eye leading to neck

Tergite: dorsal plate (used mainly in keys for abdominal segments)

Tibia (e): part of leg (see Fig. 1)

Transverse: wider than long

Trochanter: part of leg (see Fig. 1)

Ventral: lower

Vertex: area of head between and behind the compound eyes

Classification and nomenclature

Three formerly separate families are now generally accepted as subfamilies of the Staphylinidae. The Scaphidiinae were added by Kasule (1966) on the basis of similarities in larval morphology. The Pselaphinae were added by Newton & Thayer (1995) after the discovery of intermediate taxa and a cladistic analysis of adult morphological characters. More recently the Scydmaeninae were added on the basis of a cladistic analysis of larval and adult morphological characters (Grebennikov & Newton, 2009). Lawrence & Newton (1995) listed 31 extant subfamilies, now expanded to 32 after the addition of the Scydmaeninae, of which nineteen have been recorded in Britain and Ireland. Lawrence & Newton (1982) assigned subfamilies to four phylogenetic lineages (Omaliinae group, Tachyporine group, Oxyteline group and Staphylinine group). It is currently intended to publish keys to these subfamilies in the following parts:

Omaliine group

1. Omaliinae, Proteininae, Micropeplinae

2. Pselaphinae

Tachyporine group

3. Tachyporinae, Phloeocharinae, Habrocerinae, Trichophyinae

4. Aleocharinae

Oxyteline group

5. Oxytelinae, Piestinae, Scaphidiinae

Staphylinine group

6. Scydmaeninae

7. Steninae, Euaesthetinae, Oxyporinae

8. Pseudopsinae, Paederinae, Staphylininae

A checklist of species is given in each part. It is based on the most up-to-date annual version of the on-line British checklist available at **www.coleopterist.org** (latest version: 2008). Tribes and sub-tribes have been catalogued by Newton & Thayer (1992). They are not used in the keys, but they are used to organise species checklists and the sequencing of species in the main text in order to facilitate cross-referencing to standard catalogues (Löbl, 1997; Herman, 2001; Löbl & Smetana, 2004). Subgenera are treated in a similar fashion.

Key to subfamilies of Staphylinidae

1. Antennae fixed to upper surface of head so that insertion point is clearly visible from above (see Figs 2-8) 2

- Antennae fixed to the side of the front of the head so that insertion point is obscured from above by a raised or, at least definite edge (see Figs 9-23) 9

2. Antennae fixed inside the base of the mandibles (see Figs 2-4) ... 3

- Antennae fixed to the front of the head outside the base of the mandibles (see Figs 5-8) 5

3. Antennae fixed to upper surface of at or near the level of the front margins of the eyes (Fig. 2); [eyes large and bulging, occupying most of side of head] .... Steninae (2 genera, 75 spp.)

- Antennae fixed well in front of the eyes (see Figs 3, 4) 4

4. Antennae without club; head with eyes set forward (Fig. 3); pronotum without longitudinal depressions; length 3 mm or more .... Staphylininae (33 genera, 181 spp.)

- Antennae with two-segmented club; head with eyes set back toward the back of the side of the head (Fig. 4); pronotum with two longitudinal depressions; length 2 mm or less .... Euaesthetinae (1 genus, 3 spp.)

5. Elytra covering most of abdomen leaving only one or two segments exposed .... Scaphidiinae (3 genera, 5 spp.) (p. 19)

- Elytra short leaving six segments exposed 6

6. Top of head with two small simple eyes (Fig. 5) .... Omaliinae (see couplet 15)

- Simple eyes absent (Figs 6-8) 7

7. Pronotum without pubescence except for isolated setae arising from punctures; upper surface of elytra separated from reflexed sides by a sharply defined border ....Tachyporinae (tribe Mycetoporini – 7 genera, 25 spp.)

- Pronotum pubescent (sometimes sparsely, but then pronotum is very transverse); upper surface of elytra with no border separating it from smoothly curved reflexed sides 8

8. Antennae very long and thin, the apex of each segment bearing a whorl of long hairs (Fig. 7) .... Trichophyinae (1 sp.)

- Antennae short or long, but without long outstanding hairs (Fig. 8) .... Aleocharinae (part – 125 genera, 454 spp.)

9. Abdomen covered by elytra, except sometimes for tip of last segment; [antennae clubbed or thickened at apex; body length < 2.5 mm] .... Scydmaeninae (9 genera, 32 spp.)

- Abdomen with several segments visible from above 10

10. Elytra ridged with raised longitudinal keels 11

- Elytra smooth, sometimes with longitudinal rows of punctures 12

11. Antennae with nine segments, the last expanded to form a club; keels on pronotum join up to form cells (Fig. 9); keels on elytra continue to apex; keels present on abdomen; body without long hairs .... Micropeplinae (2 genera, 4 spp.)

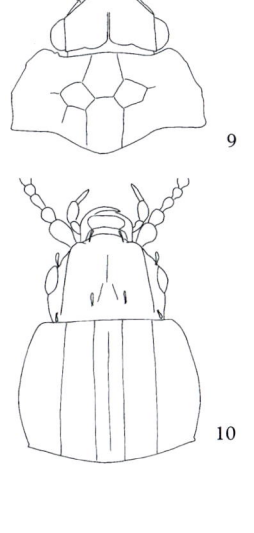

- Antennae with 11 segments; keels on pronotum straight and separate (Fig. 10); keels on elytra not continued to apex; keels absent on abdomen; body sparsely clothed with long hairs that are thickened and flattened at their tips ...... Pseudopsinae (1 sp.)

12. Abdomen inflexible with five segments visible from above; [antennae clubbed; body length < 2.5 mm] ....
.... Pselaphinae (19 genera, 53 spp.)

- Abdomen flexible with at least six segments visible from above (sometimes, if the abdomen has contracted whilst drying, fewer segments may be visible, especially in species with longer elytra) 13

13. Either head attached to pronotum by narrow neck (Fig. 11) **or** fourth tarsal segment bilobed .... Paederinae (part – 5 genera, 22 spp.)

\- Head attached to pronotum by thick neck (Fig. 12); fourth tarsal segment simple 14

14. Tarsi with three segments .... Oxytelinae (part – 11 genera, 83 spp.) (p. 23)

\- Tarsi with four or five segments 15

15. Both pronotum and head (including eyes but not including mandibles) quadrate or elongate (head rarely slightly transverse as in Fig. 13); [antennal insertion hidden by prominent raised edge] .... Paederinae (part – 9 genera, 39 spp.)

\- Either head (including eyes but not including mandibles) **or** pronotum definitely transverse (*e.g.* Fig. 14) 16

16. Simple eyes (ocelli) present on vertex of head 17

\- Simple eyes absent 18

17. Top of head with two small simple eyes, sometimes situated behind pits (Fig. 15) .... Omaliinae (28 genera, 70 spp.)

- Top of head with one central small simple eye; shape of head characteristic (Fig. 16) .... Proteininae (genus *Metopsia* – 1 sp.)

18. Body with dense clothing of golden hairs; [antennae not covered by prominent raised edge (Fig. 17); body length 2 mm or less] .... Phloeocharinae (1sp.)

- Body sparsely pubescent **or** if pubescent, then body length > 2 mm 19

19. Antennal insertion covered by prominent raised edge on side of head; head wider than or only slightly narrower than pronotum (see Figs 18-19) 20

- Antennae inserted on side of head, but not covered by prominent raised edge; head much narrower than pronotum (see Figs 20-23) 22

20. Apical segment of labial palps expanded characteristically; mandibles produced forwards and over half as long as rest of head **and** antennal segments 4 to 10 strongly transverse (Fig. 18); [colour red and black] .... Oxyporinae (1 sp.)

- Apical segment of labial palps not expanded; if mandibles produced forwards, then most antennal segments elongate (see Figs 14, 19) 21

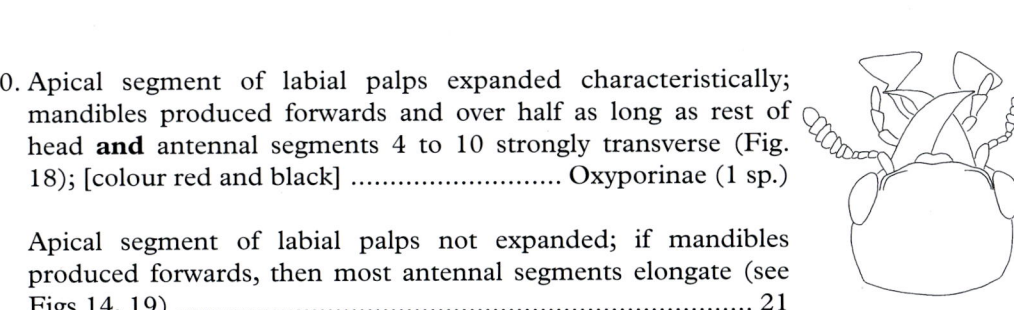

15

16

17

18

21. Elytra wider than pronotum whose surface is uneven due to depressions or furrows; horns absent ....
.... Oxytelinae (part – 4 genera, 4 spp.) (p. 23)

- Elytra and pronotum of similar width; surface of pronotum even and somewhat flattened; male armed with horns on forehead and mandibles (Fig. 19) .... Piestinae (1 sp.) (p. 22)

22. Pronotum and elytra very transverse, together approximately square in outline (Fig. 20); abdomen tapered strongly; antennae with ten segments; [body length 2 mm or less] ....
.... Aleocharinae (genus *Cypha* – 9 spp.)

- Pronotum and elytra together elongate and abdomen less tapered in insects under 2 mm; antennae with eleven segments 23

23. Antennae very long and thin, the segments and linkages combining to form a structure like a string of beads with long hairs on each segment (Fig. 21) .... Habrocerinae (1 sp.)

- Antennae often long, but without outstanding hairs 24

24. Pronotum produced forwards at sides to enclose temples (Fig. 22); raised side margins of abdomen narrow and strongly angled with surface of tergites; abdomen often more or less evenly tapered to apex; if body length less than 2.5 mm, then microsculpture feeble on upper surface ....
.... Tachyporinae (tribe Tachyporini – 7 genera, 25 spp.)

- Front margin of pronotum at most feebly produced forwards at sides (Fig. 23); raised side margins of abdomen broad and obtusely angled with surface of tergites; abdomen with rounded sides; [upper surface with strong microsculpture or strongly punctured; body length 2.5 mm or less] ....
.... Proteininae (part – 2 genera, 10 spp.)

Checklist of species
included in this volume

Family STAPHYLINIDAE Latreille, 1802
 Subfamily SCAPHIDIINAE Latreille, 1807
  Tribal classification follows Löbl (1997)
  Tribe SCAPHIDIINI Latreille, 1807
   SCAPHIDIUM Olivier, 1790
    quadrimaculatum Olivier, 1790
  Tribe SCAPHIINI Achard, 1924
   SCAPHIUM Kirby, 1837
    immaculatum (Olivier, 1790)
  Tribe SCAPHISOMATINI Casey, 1894
   SCAPHISOMA Leach, 1815
    agaricinum (Linnaeus, 1758)
    assimile Erichson, 1845
    boleti (Panzer, 1793)
 Subfamily PIESTINAE Erichson, 1839
   SIAGONIUM Kirby, 1815
    quadricorne Kirby, 1815
 Subfamily OXYTELINAE Fleming, 1821
  Tribal classification follows Herman (2001). Hansen (1996) gives an alternative classification. Gildenkov (2003) erected a new tribe, Mandini, which includes the genera *Manda* and *Planeustomus*.
  Tribe COPROPHILINI Heer, 1839
   COPROPHILUS Latreille, 1829
    ELONIUM Leach, 1819 suppressed by ICZN Op. 1722
    striatulus (Fabricius, 1792)
  Tribe DELEASTERINI Reitter, 1909
   DELEASTER Erichson, 1839
    dichrous (Gravenhorst, 1802)
   SYNTOMIUM Curtis, 1828
    aeneum (Müller, P.W.J., 1821)
  Tribe OXYTELINI Fleming, 1821
   ANOTYLUS Thomson, C.G., 1859
    clypeonitens (Pandellé, 1867)
     speculifrons: misidentified
    complanatus (Erichson, 1839)
    fairmairei (Pandellé, 1867)
    hamatus (Fairmaire & Laboulbène, 1856)
    insecatus (Gravenhorst, 1806)
    inustus (Gravenhorst, 1806)
    maritimus Thomson, C.G., 1861
     perrisi (Fauvel, 1862), synonym
    mutator (Lohse, 1963)
     sculpturatus: misidentified in part
    nitidulus (Gravenhorst, 1802)
    rugosus (Fabricius, 1775) conserved by ICZN Op. 2086
    saulcyi (Pandellé, 1867)

sculpturatus (Gravenhorst, 1806)

tetracarinatus (Block, 1799)

OXYTELUS Gravenhorst, 1802

fulvipes Erichson, 1839

laqueatus (Marsham, 1802)

migrator Fauvel, 1904

piceus (Linnaeus, 1767)

sculptus Gravenhorst, 1806

PLATYSTETHUS Mannerheim, 1830

Subgenus CRAETOPYCRUS Tottenham, 1939

alutaceus Thomson, C.G., 1861

capito Heer, 1839

cornutus (Gravenhorst, 1802)

degener Mulsant & Rey, 1878

cornutus: misidentified in part

nitens (Sahlberg, C.R., 1832)

nodifrons Mannerheim, 1830

Subgenus PLATYSTETHUS Mannerheim, 1830

arenarius (Fourcroy, 1785)

Tribe THINOBIINI Sharp, 1887

APLODERUS Stephens, 1833

HAPLODERUS error

caelatus (Gravenhorst, 1802)

BLEDIUS Leach, 1819

Subgenus BLEDIUS Leach, 1819

limicola Tottenham, 1940

germanicus Wagner, 1935, preoccupied

spectabilis Kraatz, 1857

ssp. *frisius* Lohse, 1978

tricornis (Herbst, 1784)

unicornis (Germar, 1825)

Subgenus EUCERATOBLEDIUS Znojko, 1929

furcatus (Olivier, 1811)

Subgenus ELBIDUS Mulsant & Rey, 1878

bicornis (Germar, 1822) conserved by ICZN Op. 2053

dama Motschulsky, 1857, synonym

ssp. *jutlandensis* Herman, 1986

atlanticus Lohse, 1978, preoccupied

diota Schiødte, 1866

Subgenus HESPEROPHILUS Curtis, 1829

atricapillus (Germar, 1825) conserved by ICZN Op. 2053

praetermissus Williams, 1929

atricapillus: misidentified in part by Pope (1977)

crassicollis Lacordaire, 1835

dissimilis Erichson, 1840

femoralis (Gyllenhal, 1827)

rastellus Thomson, 1867, synonym

occidentalis Bondroit, 1907

gallicus (Gravenhorst, 1806)

fracticornis (Paykull, 1790), preoccupied

laetior Mulsant & Rey, 1878, synonym

> *sharpi* Fowler, 1913, synonym
>
> *annae* Sharp, 1911
>
> *arcticus* Sahlberg, J., 1890
>
> > *denticollis*: misidentified by British authors
>
> *defensus* Fauvel, 1872
>
> > *gulielmi* Sharp, 1913, synonym
>
> *filipes* Sharp, 1911
>
> *fuscipes* Rye, 1865
>
> *longulus* Erichson, 1839
>
> *opacus* (Block, 1799)
>
> *pallipes* (Gravenhorst, 1806)
>
> > *annae*: misidentified in part by Pope (1977)
> >
> > *larseni* Hansen, 1940, synonym
>
> *terebrans* (Schiødte, 1866)
>
> *erraticus* Erichson, 1839

Subgenus ASTYCOPS Thomson, C.G., 1859

> *subterraneus* Erichson, 1839

Subgenus DICARENUS Gistel, 1834

> *fergussoni* Joy, 1912
>
> > *arenarius* (Paykull, 1800), preoccupied
> >
> > *arenoides* Tottenham, 1939, synonym
>
> *subniger* Schneider, 1900

CARPELIMUS Leach, 1819

TROGOPHLOEUS Mannerheim, 1830

Subgenus CARPELIMUS Leach, 1819

> *fuliginosus* (Gravenhorst, 1802)
>
> *lindrothi* Palm, 1942
>
> *obesus* (Kiesenwetter, 1844)
>
> *pusillus* (Gravenhorst, 1802)
>
> > *lasti* (Scheerpeltz, 1946), synonym

Subgenus MYOPINUS Scheerpeltz, 1937

> *elongatulus* (Erichson, 1839)

Subgenus PARATROGOPHLOEUS Hatch, 1957

> *bilineatus* Stephens, 1834
>
> *erichsoni* (Sharp, 1871)
>
> > *bilineatus*: misidentified in part
>
> *rivularis* (Motschulsky, 1860) conserved by ICZN Op. 2086
>
> *similis* Smetana, 1967

Subgenus TROGINUS Mulsant & Rey, 1878

> *incongruus* Steel, 1969
>
> > *zealandicus*: misidentified in part by British authors
>
> *schneideri* (Ganglbauer, 1895)
>
> > *hemerinus* (Joy, 1913), synonym
>
> *zealandicus* (Sharp, 1900)

Subgenus TROGOPHLOEUS Mannerheim, 1830

> *corticinus* (Gravenhorst, 1806)
>
> > *dispersepunctatus* (Scheerpeltz, 1947), synonym
>
> *foveolatus* (Sahlberg, C.R., 1832)
>
> *gracilis* (Mannerheim, 1830)
>
> > *graciliformis* Konzelmann & Lohse, 1981, synonym

halophilus (Kiesenwetter, 1844)
impressus (Boisduval & Lacordaire, 1835)
manchuricus (Bernhauer, 1938)
   ssp. *subtilicornis* (Roubal, 1946)
   corticinus: misidentified in part by British authors
subtilis (Erichson, 1839)
MANDA Blackwelder, 1952
  ACROGNATHUS: misidentified
  mandibularis (Gyllenhal, 1827)
OCHTHEPHILUS Mulsant & Rey, 1856
  ANCYROPHORUS Kraatz, 1858
   andalusiacus (Fagel, 1957)
    omalinus: misidentified in part by British authors
   angustior (Bernhauer, 1943)
    omalinus: misidentified in part by British authors
    venustulus: misidentified by Pope (1977)
   aureus (Fauvel, 1871)
   omalinus (Erichson, 1840)
PLANEUSTOMUS Jacquelin du Val, 1857
  flavicollis Fauvel, 1871
  palpalis (Erichson, 1839)
TEROPALPUS Solier, 1849
  CARPELIMUS: in part
  unicolor (Sharp, 1900)
   anglicanus (Sharp, 1900), synonym
THINOBIUS Kiesenwetter, 1844
  bicolor Joy, 1911
   brunneipennis: misidentified
   linderianus Scheerpeltz, 1966, synonym
   linearis: misidentified by British authors
  brevipennis Kiesenwetter, 1850
  ciliatus Kiesenwetter, 1844
   longipennis: misidentified in part
   praetor Smetana, 1959, synonym
  crinifer Smetana, 1959 conserved by ICZN Op. 2129
   longipennis: misidentified in part
   strandi Smetana, 1960, synonym
  longipennis (Heer, 1841)
  major Kraatz, 1857
   angusticeps: misidentified
   macroceros Joy, 1913, synonym
   macrocerus: misspelling
  newberyi Scheerpeltz, 1925
   pallidus Newbery, 1909, preoccupied
THINODROMUS Kraatz, 1857
  AMISAMMUS des Gozis, 1886
  CARPELIMUS: authors, in part
  arcuatus (Stephens, 1834)

10. SCAPHIDIINAE

This subfamily is unusual among the Staphylinidae in that the elytra cover most of the abdomen leaving only one or two terminal segments visible from above. They may be recognised among beetles in general by a combination of their five-segmented antennal club, free terminal abdominal segment, characteristic body shape that has been likened to a boat (a dinghy rather than a schooner) and their slender legs and antennae. All the British species have a shiny appearance and very short and sparse pubescence. Unlike other Staphylinidae, the head is held with the jaws pointing downwards rather than forwards. In the British species the eyes are kidney-shaped to varying degrees and more or less wrapped round the antennal insertion. The elytra carry impressed striae alongside the suture and the side margins. There are no obvious external differences between the sexes.

The Scaphidiinae were formerly regarded as a separate family, but incorporated into the Staphylinidae as a subfamily on grounds of larval morphology by Kasule (1966). This approach was supported by Lawrence & Newton (1982) and most subsequent authors. Newton & Thayer (1995) placed them in the oxyteline lineage of subfamilies, but Hansen (1997) considered them to be the sister-group of all other Staphylinidae.

Scaphidiinae are believed to be exclusively mycophagous (feeders on fungi) (Newton, 1984; Lipkow & Betz, 2005) and they are predominantly found in litter and fungal fruiting bodies, especially those associated with wood decay.

This subfamily contains approximately 1,300 species distributed throughout the world except for Antarctica (Löbl, 1997). The British fauna, as currently recognised, includes five species in three genera, while the Irish fauna includes two of these species in one genus.

Key to genera of Scaphidiinae

1. Antennal segments 9 and 10 elongate (Fig. 24); body length < 2.5 mm
.... 3. *Scaphisoma* (3 spp.) (p. 21)

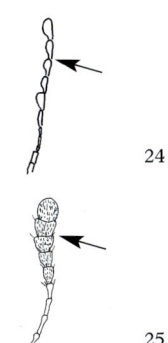

24

\- Antennal segments 9 and 10 transverse (Fig. 25); body length > 4 mm 2

25

2. Elytra with scattered punctures except for relatively strongly punctured sutural stria; each elytron with two irregular red marks .... 1. *Scaphidium* (1 sp.) (p. 20)

\- Elytra with punctured striae in addition to sutural stria; elytra all black .... 2. *Scaphium* (1 sp.) (p. 20)

1. *SCAPHIDIUM* Olivier, 1790

The base of the pronotum is strongly sinuate at sides and embraces the shoulders of the elytra. Pronotum with margin on both side and front edges and sinuate row of strong punctures near base. Tibiae with straight, longitudinal ridges, between which are rows of strong, spine-like bristles. Tergite VIII has a weakly convex apical margin. This genus contains 267 species worldwide, but only one has been recorded in Britain.

1. *Scaphidium quadrimaculatum* **Olivier, 1790** – Plate 1

Length 4.5-6 mm. Elytra black with two irregular red marks on each. Head, pronotum and apex of abdomen black. Legs black with reddish tarsi paler. Antennae reddish with black five segmented club. Pronotum covered with scattered punctures in addition to sinuate row of strong punctures near base. Elytra more strongly punctured. Sutural stria on elytra strongly punctured, continued along basal margin. Abdominal tergite VIII and part of tergite VII visible beyond apex of elytra, more weakly punctured. Pubescence on head, pronotum and terminal abdominal tergites very short and sparse, absent on elytra except for some short, upright setae.
Habitat: Associated with fungi of dead branches.
Distribution: Scattered throughout England and Wales.

2. *SCAPHIUM* Kirby, 1837

Apart from the characters mentioned in the key, this genus can be distinguished from *Scaphidium* by the rounded hind angles of the pronotum, which do not embrace the shoulders of the elytra. The sinuate row of punctures near the base of the pronotum is somewhat irregular. The rows of bristles on the tibiae are not separated by ridges. Tergite VIII has a concave apical margin. *Scaphium* species are reported to feed on fleshy Agaricales (Lipkow & Betz, 2005). The genus contains four species in the Holarctic and Afrotropical regions. Only one species has been recorded in Britain.

1. *Scaphium immaculatum* **(Olivier, 1790)** – Plate 2

Length 5-6.5 mm. Body shining black; legs pitchy black with reddish tarsi; antennae reddish with black loose five segmented club. Pronotum covered with scattered punctures in addition to sinuate row of strong punctures near base. Elytra with scattered punctures between the punctured striae, the scattered punctures being of equal strength to those on the pronotum. Strong punctures along sutural stria continued along basal margin. Abdominal tergite VIII and part of tergite VII visible beyond apex of elytra, more weakly punctured. Pubescence on head, pronotum and terminal abdominal tergites very short and sparse, absent on elytra except for some short, upright setae.
Distribution: Known to be established between 1918 and 1936 on Kent coast near St Margarets (Welch in Shirt, 1987) and rediscovered in W. Kent in 2009 (Williams, 2009).

3. *SCAPHISOMA* Leach, 1815

In this genus, the antennal club is very loose with each segment elongate, so that the antennae hardly appear clubbed at all. Pronotum with marginal bead on side edges only and no sinuate row of strong punctures near base. The sutural striae on the elytra are not punctured. The elytra are longer so that in most cases only abdominal tergite VIII is visible beyond their apices. Tergite VIII has a concave apical margin. The median lobe of the aedeagus is fairly weakly sclerotised. *Scaphisoma* contains 547 species worldwide. Three species are recorded from the Britain, two of which are also recorded from Ireland.

Key to species of *Scaphisoma*

1. Sutural striae sometimes bent outwards at base, but then only continued along base of elytra for a short distance (Fig. 27); aedeagus as in Fig. 26; [elytra black] .... 1. *agaricinum* (Linnaeus) (p. 21)

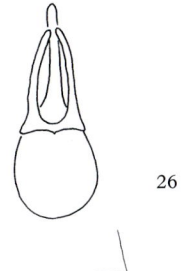

26

27

- Sutural striae continued along base of each elytron for at least one third of its breadth (Fig. 28) 2

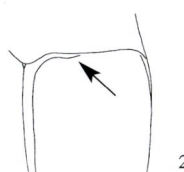

28

2. Aedeagus as in Fig. 29; elytra usually black and more densely punctured .... 2. *assimile* Erichson (p. 22)

29

- Aedeagus as in Fig. 30; elytra paler, often reddish, more sparsely punctured .... 3. *boleti* (Panzer) (p. 22)

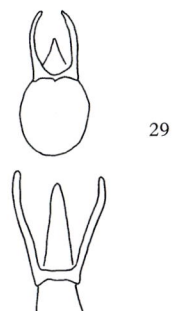

30

1. *Scaphisoma agaricinum* (Linnaeus, 1758) – Plate 3

Length 1.7-2 mm. Body black except appendages brown and pale on extreme apices of elytra and apices to abdominal tergites. Pronotum very finely punctured. Elytra more strongly punctured. Pubescence on head absent and very short on pronotum, elytra and abdominal tergite VIII.
Habitat: Found in rotten heartwood and fungal fruiting bodies associated with wood decay.
Distribution: Scattered throughout England and Wales, more rarely recorded in Scotland and Ireland.

2. *Scaphisoma assimile* Erichson, 1845

Length 1.7-2 mm. Body usually black except appendages brown and pale on extreme apices of elytra and apices to abdominal tergites.
Habitat and distribution: There are 19th Century specimens from England north to Yorkshire. The only post 1900 record comes from mouldy plant litter at Hothfield Common in Kent in 1974 (Philp, 1990).

3. *Scaphisoma boleti* (Panzer, 1793)

Length 1.7-2 mm. Body dark brown to almost reddish, elytra not paler at extreme apices appendages pale, abdominal tergite VIII yellow. Punctures and pubescence similar to *S. agaricinum*.
Habitat: Found in rotten heartwood and fungal fruiting bodies associated with wood decay.
Distribution: Scattered throughout most of England and Wales, more rarely recorded in northern England, Scotland and Ireland.

11. PIESTINAE

It is possible that most species of Piestinae are saprophagous (Lipkow & Betz, 2005). The subfamily contains 107 species in 8 genera distributed throughout the world except for the Afrotropical region. Only one species occurs in Britain and Ireland.

1. SIAGONIUM Kirby, 1815

Males with well developed sexual characters are easily recognised by the horns on the head and their expanded mandibles. This genus contains 22 species from the Holarctic and Neotropical regions.

1. *Siagonium quadricorne* Kirby, 1815 – Plate 4

Length 4-5.5 mm. Body generally flattened, brown to pitchy, but elytra and sometimes pronotum paler and almost reddish; apex of elytra sometimes darker; all appendages reddish. Upper surface shining between punctures except for strong cellular microsculpture on neck and abdomen (especially tergite III). Weaker cellular microsculpture on clypeus and male horns. Weak longitudinal microsculpture on sides of pronotum. Punctures of two different diameters on pronotum and arranged in irregular striae on elytra. Pubescence on head, pronotum and elytra absent except for short marginal setae. Clothing of short pubescence on abdominal tergites in addition to numerous long, erect setae. Two rows of spines on mid tibiae; one row of spines on front tibiae. All antennal segments elongate, more so in males, segments 3 to 11 with long outstanding hairs. Males armed with expanded mandibles and forward-projecting horns on head covering antennal insertions; the development of secondary sexual characters varies markedly between individuals.
Habitat: Under bark of deciduous trees.
Distribution: England, Wales and Ireland.
Biology: The developmental history is described briefly by Beaver (1967). Both larval and adult gut contents have been reported to consist largely of fungal hyphae and yeast material mixed with remains of decaying cambium (Crowson, 1982).

12. OXYTELINAE

This diverse group characteristically has a transverse, more or less rectangular flattish head with the antennae fixed to the side of the head under a raised edge or tubercle, reminiscent of many Omaliinae. However, unlike Omaliinae, Oxytelinae lack ocelli. Some genera with a quadrate head or elongate pronotum could be mistaken for members of the Paederinae, but they can always be separated on an underside character. In the Oxytelinae each hind coxae is transverse, the base extending along the margin of the metasternum. In several genera an additional tergite (II) is visible at the base of the abdomen. All but four British genera have three-segmented tarsi. The surface of the pronotum is frequently marked by furrows and depressions whose patterns can be diagnostic for individual genera. The hairs on the head and pronotum are normally directed inwards and forwards on the front of the head. There is a tendency in several genera for the fifth and sometimes the seventh antennal segments to be longer than the fourth and sixth. The males often sport secondary sexual characters, most dramatically in the genus *Bledius*, some of which are armed with horns on the head and pronotum. Some sexual characters on the abdominal sternites are important for species identification and this should be taken into account when carding reference specimens. The male genitalia are also useful for species identification, but, to date, there have been very few references in the literature to the use of female genitalia for this purpose.

A wide range of both specialised and general habitats are exhibited by Oxytelinae. Several genera are wetland or riparian and some, most notably *Bledius*, have body forms that appear to be morphologically adapted for burrowing into soft sediments. Other genera are predominantly associated with litter or dung. A variety of chemicals are produced from abdominal glands to deter predators (Dettner & Schwinger, 1982; Steidle & Dettner, 1995a) (see also citations given under individual genera and species).

This subfamily contains 2,008 species in 48 genera distributed throughout the world except for Antarctica. The British fauna, as currently recognised, includes 91 species in 15 genera. In Ireland 48 of these species in 13 genera have been recorded.

Key to genera of Oxytelinae

1. Pronotum with central longitudinal furrow or channel longer than half its length (see Figs 31-34) 2

- Central longitudinal furrow on pronotum absent or shorter than half its length (see Figs 35, 38-41) 6

2. Depressions on pronotum represented solely by a narrow central channel (see Figs 31-32) 3

- Pronotum with other depressions as well as central furrow (see Figs 33-34) 4

3. Temples more or less absent; eyes very convex and projecting out sharply from the side of the head (Fig. 31); front tibiae with two rows of spines; pronotum and elytra more arched with narrow waist between them .... 8. *Bledius* (part – 21spp.) (p. 43)

- Temples rounded; eyes flatter and hardly projecting from the side of the head (Fig. 32); front tibiae with one row of spines; pronotum and elytra flatter with waist less obvious .... 6. *Platystethus* (7 spp.) (p. 39)

4. Elytra with punctures arranged into striae which are separated by convex intervals; central furrow on pronotum ending well before base where it is replaced by a ridge between two oblique, oval, basal depressions (Fig. 33) .... 1. *Coprophilus* (1 sp.) (p. 26)

- Elytra with punctures scattered; pronotum with two linear depressions parallel to central furrow, which stretches almost to base (Fig. 34) 5

5. Scutellum heart-shaped with a central longitudinal keel .... 4. *Anotylus* (13 spp.) (p. 28)

- Scutellum flat and diamond-shaped .... 5. *Oxytelus* (5 spp.) (p. 37)

   Note: In many carded specimens most of the scutellum is hidden under the hind margin of the pronotum, so the species keys below address these two genera together.

6. Side margin of pronotum reflexed underneath at least in front half and not visible from above 7

- Side margin of pronotum completely visible from above 10

7. Pronotum with a deep, curved, transverse basal depression which is not interrupted by a central ridge (Fig. 35) .... 15. *Thinodromus* (1 sp.) (p. 82)

- Basal depressions, if present, separated by a central ridge 8

8. Sutural angles of elytra rounded leaving a gap stretching one quarter of the length of the suture (Fig. 36); pronotum more transverse and less narrowed towards base; [body length < 2.5 mm] .... 14. *Thinobius* (7 spp.) (p. 77)

- Sutural angles of elytra sharper leaving a much smaller gap when the elytra are closed (Fig. 37); pronotum less transverse and clearly more narrowed towards base than towards front; [body length > 1.5 mm] 9

9. Elytra elongate and much wider than pronotum; pubescence on elytra dense; [body length. > 3 mm] ... 13. *Teropalpus* (1 sp.) (p. 77)

- If elytra elongate, then barely wider than pronotum; pubescence sparser .... 9. *Carpelimus* (17 spp.) (p. 60)

10. Elytra with punctured striae and/or longitudinal ridges 11

- Elytra with punctures scattered and irregular; ridges on elytra absent 12

11. Elytra weakly punctured between unpunctured ridges, any lines of punctures being short or irregular; body length > 4 mm .... 10. *Manda* (1 sp.) (p. 73)

- Front of elytra with strongly punctured striae; body length < 3 mm .... 12. *Planeustomus* (2 spp.) (p. 76)

12. Elytra roughly twice as long as pronotum 13

- Elytra clearly less than twice as long as pronotum 14

13. Head and abdomen black contrasting with red-brown pronotum and elytra; body length 6.5-7 mm; head with broad oblique furrows stretching back from inside margin of eyes; pronotum with basal depression uninterrupted by central ridge (Fig. 38) .... 2. *Deleaster* (1 sp.) (p. 27)

- Colour pattern different; body length < 6mm; furrows absent on back of head; basal depressions on pronotum, if present, separated by central ridge (Fig. 39) .... 11. *Ochthephilus* (4 spp.) (p. 73)

14. Elytra transverse, barely longer than pronotum; pronotum with side margins crenulate and hind angles toothed (Fig. 40); head and pronotum strongly and densely punctured but shiny and metallic .... 3. *Syntomium* (1 sp.) (p. 27)

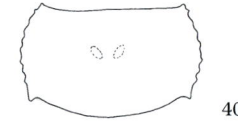

40

- Elytra quadrate to elongate; pronotum with side margins smooth and hind angles rounded (see Fig. 41); head and pronotum finely or more sparsely punctured and matt through surface sculpture 15

15. Pronotum transverse with depressions either side of the centre line (Fig. 41); front tibiae with just one row of spines; body flatter .... 7. *Aploderus* (1 sp.) (p. 43)

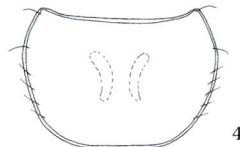

41

- Pronotum quadrate to elongate, lacking depressions except sometimes for abbreviated central channel; front tibiae with two rows of spines; body form more cylindrical .... 8. *Bledius* (part – 4 spp.) (p. 43)

1. *COPROPHILUS* Latreille, 1829

A distinctive genus within the Oxytelinae. The shape of the pronotum, the serial punctures on the elytra, the small convex eyes and the absence of rounded temples distinguish it from the genera *Oxytelus* and *Anotylus*. There are five tarsal segments. 30 species of *Coprophilus* are distributed throughout the Holarctic region. Only one species has been recorded in Britain and Ireland.

1. *Coprophilus striatulus* (Fabricius, 1792) – Plate 5

Length 6.5-7 mm. Body black, with reddish margins to the pronotum, elytra and abdominal tergites. Legs and antennae dark red to black, tarsi red. Upper surface shining between the punctures except for abdominal tergites, which have strong cellular microsculpture. Antennae thick, segments 4 to 10 roughly quadrate. Eyes convex but much smaller than temples which are straight and parallel-sided. Punctures on head becoming much stronger and denser toward base. Head and pronotum with long, curved, erect marginal setae. Pronotum flat and quadrate with weakly crenulate sides. Elytra with irregular longitudinal striations along basal margin behind punctured striae. Punctures on abdominal tergites directed backward, bearing semi-erect hairs and becoming sparser toward apex of abdomen. Tergite VIII with concave hind margin.
Habitat: Associated with a wide range of wetland and riparian environments.
Distribution: Widely recorded from East Anglia, but also scattered through much of England, S. Wales, Scotland and N. Ireland.
Biology: Active mainly in early spring when sometimes recorded in dispersal flights. The defensive secretions from glands on the abdominal tergites have been studied by Dettner (1983).

2. *DELEASTER* Erichson, 1839

Strikingly distinctive within the Oxytelinae, *Deleaster* more closely resembles a large *Anthophagus* from the Omaliinae, by virtue of its long legs, wide raised abdominal margins and small pronotum compared to the long, wide elytra. However, it can be easily recognised by the depressions on the head and pronotum. There are five tarsal segments. This genus contains around ten species from the Holarctic and Afrotropical regions. Only one species has been recorded in Britain and Ireland.

1. *Deleaster dichrous* (Gravenhorst, 1802) – Plate 6

Length 6.5-7 mm. Head and abdomen black, pronotum and elytra chestnut brown; legs and antennae red. All antennal segments clearly elongate. Head with diagonally slanted furrows inside eyes. Eyes larger than scarcely rounded temples, angle between hind margin of eyes and temples very shallow. Head and pronotum with semi-erect, yellow pubescence, both shiny toward front with non-cellular microsculpture and matt toward rear with strong cellular microsculpture. Elytra depressed directly behind scutellum and less markedly along suture, with somewhat dense pubescence, hairs directed more or less backwards except around scutellum and along hind margin. Abdominal tergites with transversely cellular microsculpture and somewhat dense pubescence. Tergite VIII with deep, irregular incisions in middle of slightly convex apical margin.

Habitat: Associated with a wide range of exposed substrates by streams which are often partly shaded. Also reported from silage clamps (Anderson, 1986).

Distribution: Widely recorded in SW. England and S. Wales, and locally distributed throughout Britain and Ireland.

Biology: Attracted to light traps (J.G. Woodhead, pers. comm.). The defensive secretions from glands on the abdominal tergites have been studied by Dettner *et al.* (1985).

3. *SYNTOMIUM* Curtis, 1828

Very distinctive within the Oxytelinae and unlikely to be confused with any other genus. The abdomen is wide with wide raised margins and reminiscent of Omaliinae, but the elytra are strongly transverse, unlike most Omaliinae. There are five tarsal segments. This genus contains eight species, all restricted to the Holarctic region. Only one species has been recorded in Britain and Ireland.

1. *Syntomium aeneum* (Muller, 1821) – Plate 7

Length 2.2-3 mm. Body black with metallic bronze-green sheen; legs dark red. Antennae with three-segmented club, apical segments transverse and paler than dark red basal segments. Abdomen with cellular microsculpture contrasting with rest of body which is strongly punctured and shining. Pubescence on whole upper surface short and erect. Eyes much longer than well-rounded temples. Sides of pronotum crenulate, hind angles produced into a strong tooth and hind margin strongly sinuate at sides. Abdominal tergite VIII with an almost triangular concavity in the hind margin.

Habitat: Ground-living, on dry, mainly nutrient-poor soils, especially in upland areas.

Distribution: Widespread in Britain and Ireland, though absent or rare in some lowland districts.

4 and 5. ANOTYLUS and OXYTELUS

Oxytelus is superficially similar to *Anotylus,* but scutellum diamond-shaped and first antennal segment never triangular in outline. In many carded specimens most of the scutellum is hidden under the hind margin of the pronotum, so the species keys below address these two genera together.

4. ANOTYLUS Thomson, 1859

Wide, fairly flat insects with short legs. Scutellum heart-shaped with central keel. Antennal segment 1 expanded toward apex and almost triangular in outline. Hind angle of pronotum obtuse, but always well marked unless obscured by crenulations (*A. rugosus*). Back of head, pronotum and elytra often with irregular, coarse longitudinal striations making the surface appear rough. Abdominal tergites usually with cellular microsculpture. Pubescence on head, pronotum and elytra normally absent except for marginal setae, and generally short on abdominal tergites. Apical margin of abdominal tergite VIII weakly concave revealing tergite IX. Males have modifications to abdominal sternites VI to VIII. In many species, they also have larger heads with expanded temples making the eyes look smaller, although this last sexual character varies enormously between individuals. Formerly included in the genus *Oxytelus*, there are 355 species worldwide, of which 13 have been recorded in Britain and eight in Ireland. A further, unnamed species awaits formal description.

Key to species of *Anotylus* and *Oxytelus*

1. Side margins of pronotum crenulate (Fig. 42) 2

- Side margins of pronotum smooth except sometimes for some very feeble crenulations toward hind angle 4

2. Antennal segment 1 red and shiny with a more or less constant diameter along most of its length; abdominal sternite VIII of male modified as in Fig. 58; [elytra black or dark brown; clypeus shiny, but with weak cellular microsculpture] ... 5.1. *O. fulvipes* Erichson (p. 37)

- Antennal segment 1 expanded toward apex, microsculptured and black or darker than segment 2 3

3. Clypeus matt due to cellular microsculpture, striations more or less extending across vertex (Fig. 43); abdominal sternites VI to VIII of male characteristic (see Fig. 58) ....
.... 4.10. *A. rugosus* (Fabricius) (p. 35)

- Clypeus shiny and lacking cellular microsculpture, striations on vertex less extensive (Fig. 44); abdominal sternites VI to VIII of male characteristic (see Fig. 58); [elytra red or red-brown] ....
.... 4.5. *A. insecatus* (Gravenhorst) (p. 34)

4. Antennal segment 1 constricted before apex; (see Figs 45-46) ... 5

- Antennal segment 1 without constriction before apex (see Figs 47-51, 56-57) 6

5. Females with large eyes, but temples well rounded (Fig. 45); abdominal sternite VIII of male characteristic (see Fig. 58) .... 5.2. *O. laqueatus* (Marsham) (p. 37)

45

- Females with even larger eyes, temples reduced and not rounded (Fig. 46); abdominal sternite VIII of male characteristic (see Fig. 58) .... 5.4. *O. piceus* (Linnaeus) (p. 38)

46

6. Centre of clypeus lacking cellular microsculpture and usually shiny 7

- Clypeus matt due to cellular microsculpture 11

7. Eyes large, occupying more than two thirds of side of head; temples absent (Fig. 47); abdominal sternite VIII of male characteristic (see Fig. 59) .... 5.3. *O. migrator* Fauvel (p. 37)

47

- Eyes occupying less than half of side of head; temples well-developed (see Figs 48-49) 8

8. Cellular microsculpture extending over whole of vertex on head (Fig. 48); abdominal sternite VIII of male characteristic (see Fig. 59); [body length < 2.5 mm] .... 4.1. *A. clypeonitens* (Pandellé) (p. 33)

48

- Cellular microsculpture absent or confined to patch inside antennal tubercles (Fig. 49) 9

49

9. Clypeus as strongly punctured as on vertex of head, although the punctures are sparser; body length < 3 mm; [elytra usually light brown] .... 4.9. *A. nitidulus* (Gravenhorst) (p. 35)

- Punctures on clypeus smaller and sparser than on vertex of head (as in Fig. 49) or even absent; body length > 3 mm 10

10. Antennal segments predominantly black; pubescence on pronotum, elytra and abdominal tergites shorter ....
    .... 4.6. *A. inustus* (Gravenhorst) (p. 35)

- Antennal segments 2-11 red; pubescence on pronotum, elytra and abdominal tergites longer ...... 4.7. *A. maritimus* Thomson (p. 35)

11. Eyes large, occupying more than two thirds of side of head (Fig. 50); temples reduced and not rounded ....
    .... 5.5. *O. sculptus* Gravenhorst (p. 38)

- Eyes occupying less than half of side of head (see Fig. 51); temples well-developed and rounded 12

50

12. Cellular microsculpture absent from vertex (Fig. 51) and on pronotum confined to mid channel; body length > 3.5 mm 13

- Strong cellular microsculpture present on most of vertex and pronotum (see Figs 56, 57); body length < 3.5 mm 14

51

13. Parameres of aedeagus expanded at apex, but not bent (Fig. 52); abdominal sternite VII in male with two tubercles near the hind margin (see Fig. 58) clothed with a dusting of dense, very short yellow pubescence extending into a depression between the tubercles .... 4.12. *A. sculpturatus* (Gravenhorst) (p. 36)

52

- Parameres of aedeagus bent just before apex but not expanded (Fig. 53); abdominal sternite VII in male with two tubercles near the hind margin (see Fig. 58) clothed with short discrete hairs ....
    .... 4.8. *A. mutator* (Lohse) (p. 35)

53

14. Outer edge of front tibiae straight and evenly curved at apex (Fig. 54) .... 4.13. *A. tetracarinatus* (Block) (p. 36)

54

- Outer edge of front tibiae strongly indented at apex (Fig. 55) ....
.... 15

55

15. Keel inside eye continued beyond back of eye (Fig. 56); body length > 2.5 mm; male abdominal sternites as in Fig. 58 ....
.... 4.2. *A. complanatus* (Erichson) (p. 34)

56

- Keel inside eye not continued beyond mid-point of eye (Fig. 57); body length < 2.5 mm 16

57

16. Tergites sparsely punctured and shining with no microsculpture visible at 60x magnification; male abdominal sternite VII with two tubercles inside hind margin which is simple (see Fig. 59) ....
.... 4.3. *A. fairmairei* (Pandellé) (p. 34)

- Tergites with microsculpture connecting sparse punctures 17

17. Hind margin of male abdominal sternite VII produced into an inwardly curving spine (see Fig. 59); [body length < 1.8 mm] .....
.... 4.4. *A. hamatus* (Fairmaire & Laboulbène) (p. 34)

- Hind margin of male abdominal sternite VII produced into a wide rectangular plate which frames a small tubercle (see Fig. 59); [body length > 1.7 mm] .... 4.11. *A. saulcyi* (Pandellé) (p. 36)

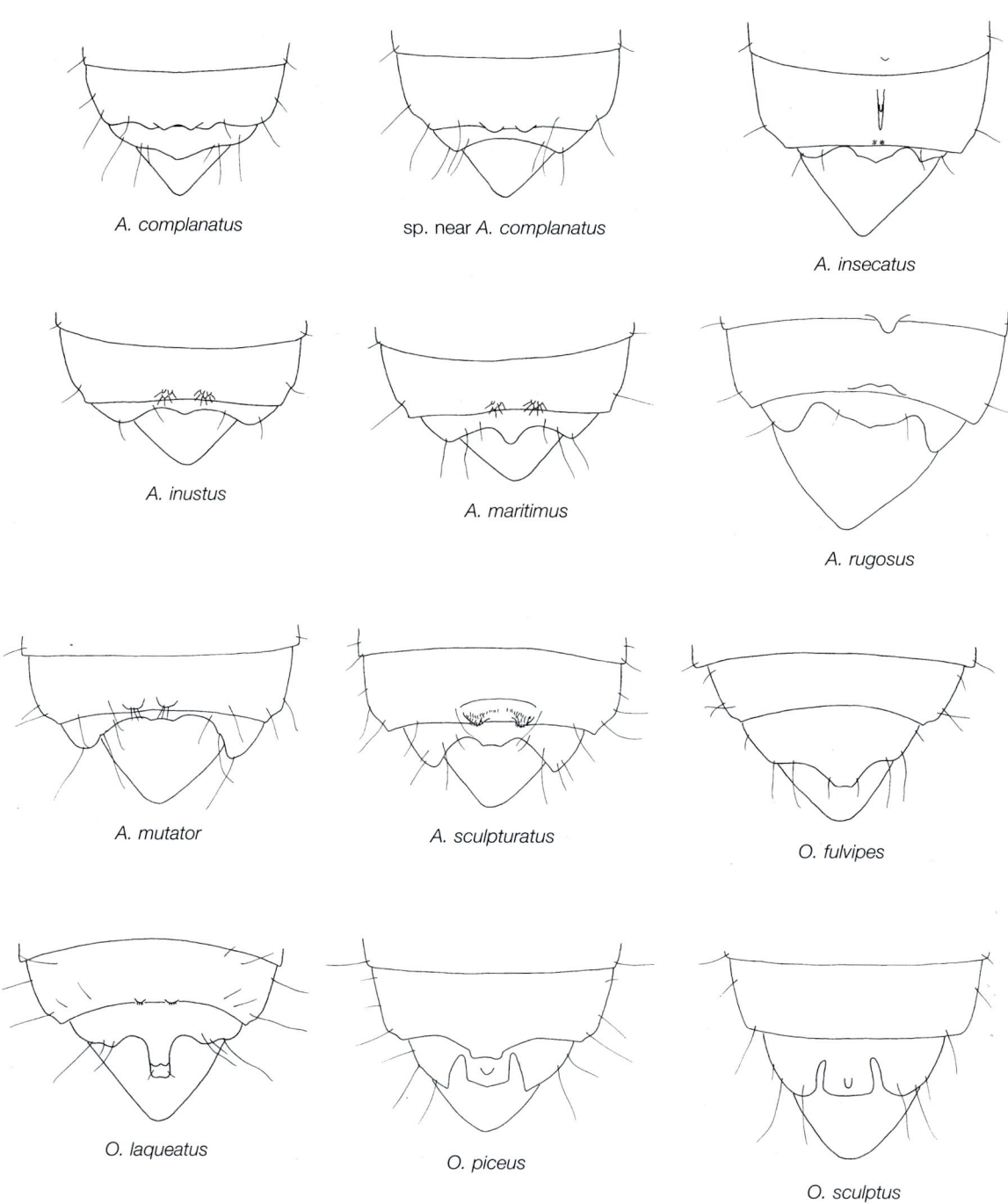

A. complanatus

sp. near A. complanatus

A. insecatus

A. inustus

A. maritimus

A. rugosus

A. mutator

A. sculpturatus

O. fulvipes

O. laqueatus

O. piceus

O. sculptus

Figure 58. Terminal abdominal sternites of males of larger species of *Anotylus* and *Oxytelus* (> 3 mm).

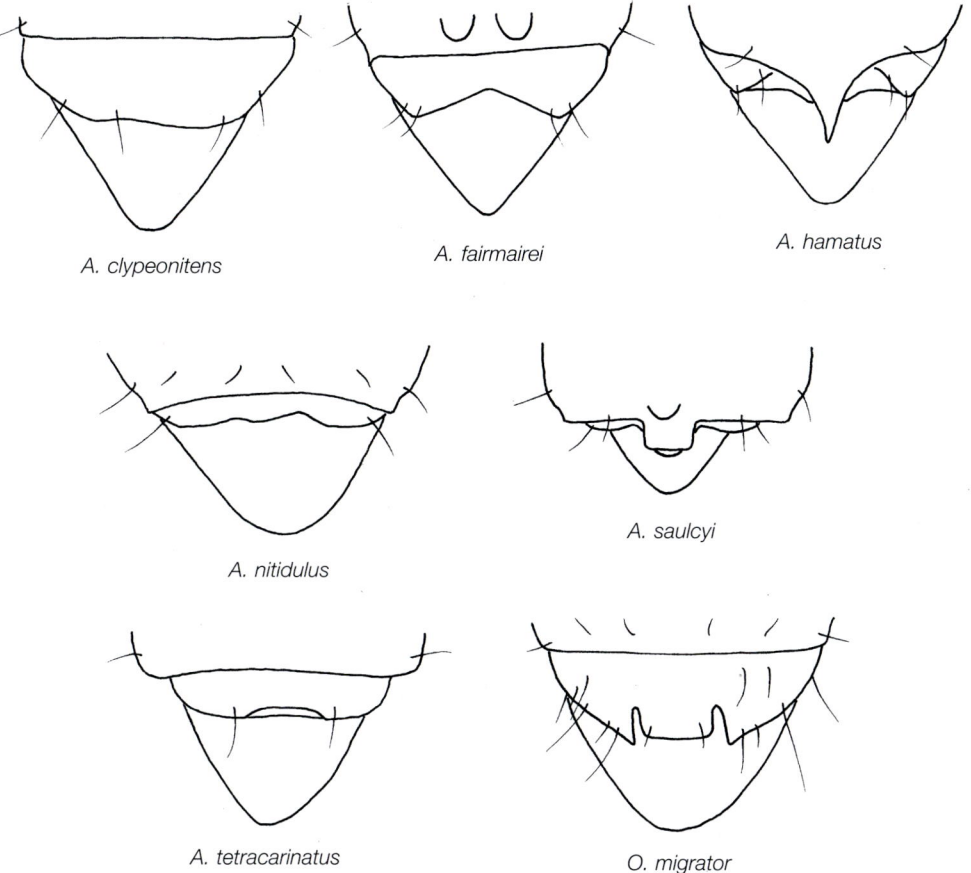

Figure 59. Terminal abdominal sternites of males of smaller species of *Anotylus* and *Oxytelus* (< 3 mm).

1. *Anotylus clypeonitens* (Pandellé, 1867)

Length 2-2.5 mm. Predominantly matt black with brown elytra and yellow legs. Ridges on pronotum shiny contrasting with matt background. Punctures on elytra visible as shiny black spots on a matt background. Abdominal tergites with relatively sparse microsculpture connecting the punctures and hence relatively shiny. Males with sternite VII unmodified and sternite VIII asymmetric (see Fig. 59).
Similar species: Similar in size to *A. tetracarinatus* and the other smaller species, but separable on the completely shiny clypeus and male secondary sexual characters. Also, elytra longer than in species of similar size.
Habitat: Compost heaps.
Distribution: England north to the Midlands, but recent records are few away from SE. England; possibly declining.

2. *Anotylus complanatus* (Erichson, 1839)

Length 3-3.5 mm. Body and antennae matt black, elytra often dark brown. Legs yellow, femora often darkened. Antennae black. Microsculpture on head more striate toward the rear, weaker on antennal tubercles and absent from two narrow, shiny, longitudinal stripes either side of the clypeus. Punctures on elytra visible as shiny black spots on a matt background. Two British species are currently confused under this name (P.M. Hammond, pers. comm.). The true *A. complanatus* has shiny ridges among the matt background of the pronotum, while in a second, undescribed species these ridges are matt with strong microsculpture. There are also differences in the sexual characters on abdominal sternites VII and VIII (see Fig. 58) and the aedeagus.

Similar species: Like a large *A. tetracarinatus*; smaller and much duller than *A. sculpturatus*.

Habitat: Dung and litter.

Distribution: Found throughout Britain and Ireland.

3. *Anotylus fairmairei* (Pandellé, 1867)

Length 1.7-2.5 mm. Predominantly black with dark brown elytra and brown legs. Forebody predominantly matt, abdominal tergites shining.

Similar species: Similar to *A. tetracarinatus,* and the other smaller species, but distinguishable on abdominal characters: males have two large tubercles on sternite VII and a V-shaped incision on sternite VIII (see Fig. 59); also separated from *A. clypeonitens*, *A. hamatus* and *A. saulcyi* on the lack of microsculpture on the tergites and slightly closer pubescence; separated from *A. tetracarinatus* on the tergite punctures which are slanted toward the rear, but not tuberculate.

Habitat: Dung.

Distribution: Formerly widespread in England and Scotland, but it was last recorded in 1947 (Hyman, 1994) and it has obviously undergone a drastic decline.

4. *Anotylus hamatus* (Fairmaire & Laboulbène, 1856)

Length 1.2-1.8 mm. Predominantly matt black with yellow-brown legs.

Similar species: Slightly smaller than *A. tetracarinatus*, *A. fairmairei* and *A. saulcyi*. Also the head lacks any shiny areas at all and the punctures on the elytra are scarcely visible against the densely sculptured background. Males have a prominent inward-curving spine on abdominal sternite VII and a highly modified sternite VIII (see Fig. 59).

Habitat: Dung.

Distribution: SE. and SW. England.

5. *Anotylus insecatus* (Gravenhorst, 1806)

Length 4-5 mm. Body brown to black with red elytra and brown pronotum. Legs red. Antennae generally dark with segments 2 to 4 reddish. Head shiny, completely lacking cellular microsculpture. Longitudinal striations on head confined to sides of vertex, absent from pronotum and sparse on elytra. Males with a small tubercle on abdominal sternite VI and a keel on sternite VII (see Fig. 58).

Similar species: Most easily confused with the form of *A. rugosus* with red elytra, but distinguished by smaller eyes and more shiny head and abdomen. In addition the side margins of the pronotum are less evenly rounded, being somewhat straightened in the basal half.

Habitat: Uncertain; recorded in a wide variety of situations.

Distribution: SE. England north to the Midlands, N. Wales and SW. Ireland, but recent records are confined to the eastern half of its range; possibly declining.

6. *Anotylus inustus* (Gravenhorst, 1806)

Length 3-4 mm. Body and antennae black, elytra occasionally light brown. Femora brown, tibiae and tarsi yellow. Head with punctures of variable size and very weak longitudinal striations, patch of cellular microsculpture inside antennal tubercles sometimes present. Males with two tubercles on abdominal sternite VII (see Fig. 58).
Similar species: Rather like a shiny *A. sculpturatus*.
Habitat: Dung and litter on dry ground, preferring insolated areas of bare ground and short turf.
Distribution: Mainly lowland areas of Britain and Ireland and much more scattered in the north and west.

7. *Anotylus maritimus* Thomson, 1861

Length 3-3.8 mm. Body black with yellow to brown elytra; pronotum sometimes brown. Legs yellow. Antennae red with basal segment slightly darker, short with segments 5 to 10 transverse. Head always with cellular microsculpture inside antennal tubercles. Longitudinal striations weak on head, pronotum and elytra. Yellow pubescence visible, although sparse, on elytra and especially noticeable on abdomen. Male abdominal sternites similar to *A. inustus*.
Similar species: Separable from *A. inustus* on antennal colour and longer pubescence.
Habitat: Litter and dung on sandy shores.
Distribution: Widespread around the coasts of Britain and Ireland, but more frequently encountered on western coasts.

8. *Anotylus mutator* (Lohse, 1963)

Length 3.5-4.5 mm.
Similar species: Only separable from *A. sculpturatus* on male sexual characters.
Habitat: Dung, carrion and marsh litter.
Distribution: First recognised as British by Hammond (1968). Widespread in England and S. Scotland, but much more rarely encountered than *A. sculpturatus*.

9. *Anotylus nitidulus* (Gravenhorst, 1802)

Length 2.3-3 mm. Predominantly shiny black except elytra yellow to brown and legs red. Keel evident inside eye. Abdominal pubescence short and sparse. Males with unmodified abdominal sternite VII and a somewhat irregular, truncate apical margin to sternite VIII (see Fig. 59).
Similar species: Resembles a small *A. inustus* with pale elytra.
Habitat: Compost and dung.
Distribution: Formerly widespread in Britain and E. Ireland, but there are relatively few recent records and it has apparently undergone a drastic decline in many areas.
Biology: Disperses by flight during the evening (Omer-Cooper & Tottenham, 1934; Twinn, 1958).

10. *Anotylus rugosus* (Fabricius, 1775) – Plate 8

Length 4.5-5 mm. Body black with paler appendages, elytra occasionally bright red, more rarely pronotum, elytra and abdomen yellow to brown. Antennae long with most segments elongate. Size of head variable, but usually only slightly narrower than pronotum, usually widest behind the eyes, which are often small. Sides of pronotum more strongly crenulate than in any other species of *Anotylus* or *Oxytelus* and more evenly rounded than other species with crenulate margins. Longitudinal striations fairly strong and coarse on head, but weak on pronotum and elytra, which are consequently shiny. Abdominal tergites relatively matt due to particularly strong cellular microsculpture. Males with a prominent tubercle on abdominal sternite VI (see Fig. 58).

Habitat: Damp vegetable litter, especially in marshes.

Distribution: Widespread throughout the British Isles and one of the most commonly encountered *Anotylus* species.

Biology: Taken in flight at sunset (Omer-Cooper & Tottenham, 1934). Also recorded at light (Lane & Mann, 2006).

11. *Anotylus saulcyi* (Pandellé, 1867)

Length 1.7-2.5 mm. Body and antennae matt black. Legs brown.

Similar species: Similar in size to *A. tetracarinatus,* but, apart from tibial character, microsculpture on abdominal tergites evident and punctures on elytra relatively dense and contrasting less with matt background. Males with a tubercle on abdominal sternite VII and a rectangular projection on the apical margin (see Fig. 59).

Habitat: Underground mammal nests.

Distribution: England north to Yorkshire, but recently recorded only from eastern areas.

12. *Anotylus scuplturatus* (Gravenhorst, 1806) – Plate 9

Length 3.5-4.5 mm. Body black with yellow legs. Femora slightly darkened. Elytra sometimes dark brown. Antennae dark and short with segments 5 to 10 transverse. Cellular microsculpture on both clypeus and frontal areas of head, but coarser on frons. Vertex with striations and central furrow. Head usually wider than pronotum in males. Pronotum punctured and with striations except on ridges, which are shiny.

Habitat: Dung and vegetable litter.

Distribution: Widespread in the British Isles and one of the most commonly encountered *Anotylus* species.

Biology: Larvae have been observed feeding directly on dung (Hinton, 1944). Taken in flight at sunset (Omer-Cooper & Tottenham, 1934) and in light traps (Welch, 1977).

13. *Anotylus tetracarinatus* (Block, 1799)

Length 1.8-2.5 mm. Body completely matt black with yellow legs, femora slightly darkened. Antennae dark and short with segments 5 to 10 transverse. Head completely matt except for narrow crescent-shaped, shiny area framing the clypeus. Ridges on pronotum matt. Cellular microsculpture on abdominal tergites extremely weak, but surface appearing wrinkled at 30x magnification because of the slightly tuberculate form of the punctures, which are directed backward. Males with modifications to abdominal sternite VIII, but not sternite VII (see Fig. 59).

Similar species: Similar in size to *A. clypeonitens, A. fairmairei, A. hamatus* and *A. saulcyi,* but separable on the lack of an apical incision on the front tibiae.

Habitat: Dung on dry ground. This species is a prolific disperser and individuals can turn up almost anywhere.

Distribution: Widespread in the British Isles and one of the most commonly encountered *Anotylus* species; much more abundant everywhere than other *Anotylus* species of a similar size.

Biology: Taken in flight at sunset (Omer-Cooper & Tottenham, 1934). Also recorded at light (Welch, 1977; Lane & Mann, 2006).

5. *OXYTELUS* Gravenhorst, 1806

Superficially similar to *Anotylus*, but scutellum diamond-shaped and first antennal segment never triangular in outline. A key to the species of both genera is included on page 28. Sexual differences as in *Anotylus*. Worldwide this genus contains 196 species, of which five have been recorded in Britain. Two of these species are known from Ireland.

1. *Oxytelus fulvipes* Erichson, 1839 – Plate 10

Length 4-4.8 mm. Body generally shiny, black with elytra slightly paler. Legs yellow. Antennae bicoloured with sharp change in colour between segments 4 and 5. This change is associated with a change in surface sculpture pubescence, segments 5 to 11 being matter and more hairy. Front of head shiny due to weak cellular microsculpture, back of head shiny with strong, round punctures. Longitudinal striations virtually absent except on elytra. Sides of pronotum only weakly crenulate, straightened toward base or even vaguely sinuate (see Fig. 42). Elytra with very short, sparse pubescence. Cellular microsculpture on abdominal tergites relatively strong.
Habitat: Undisturbed marsh with fluctuating water levels such as carr, shaded ditches and oxbow lakes in river floodplains.
Distribution: Widespread in England and S. Wales, but rarely encountered.

2. *Oxytelus laqueatus* (**Marsham, 1802**) – Plate 11

Length 4-4.5 mm. Body black with elytra (except for base) and legs yellow. Antennae bicoloured and short, segments 5-10 transverse, segments 5 to 11 matter and more hairy than segments 1 to 4. Front margin of clypeus toothed in males and obtusely angled in females with central section concave or straight (see Fig. 45). Clypeus shiny, lacking microsculpture. Two small longitudinal furrows at back of head plus a central furrow which may be weak or absent. Longitudinal striations absent except on elytra, where they are very weak. Males with narrow projection on abdominal tergite VIII (see Fig. 58).
Habitat: Dung.
Distribution: Widespread in Britain and Ireland. One of the most commonly encountered Staphylinidae in dung.
Biology: Taken in flight at sunset (Omer-Cooper & Tottenham, 1934). Also recorded at light (Lane & Mann, 2006).

3. *Oxytelus migrator* Fauvel, 1904

Length 2.5-3 mm. Predominantly yellow-brown with darker brown head and pronotum. Legs and antennae yellow. Head triangular and smaller than pronotum. Punctures and striations on head rather ill-defined. Neck with sparse non-cellular microsculpture. Pronotum, elytra and abdominal tergites with long, upright setae. Pronotum sparsely punctured with striations restricted to sides. Males with two narrow incisions on the apical margin of sternite VIII (see Fig. 59).
Similar species: Resembles a miniature *O. sculptus*, but with even larger eyes relative to the rest of head, no cellular microsculpture on head and a concave front margin to the clypeus between two obtuse angles.
Habitat: Likely to turn up in compost heaps and dung (Hammond, 1998a).
Distribution: Recent immigrant reported from Surrey, London, Wiltshire, Warwickshire and Liverpool, but undoubtedly more widely distributed.
Biology: Recorded at light (Lane & Mann, 2006) and in a flight interception trap (Owen, 1997).

4. *Oxytelus piceus* (Linnaeus, 1767)

Length 4-5 mm. Coloration as in *O. laqueatus*.

Similar species: Closely resembles *O. laqueatus*, but males with a wider projection on abdominal sternite VIII (see Fig. 58), tergites duller due to stronger microsculpture, only one central longitudinal furrow at back of head, clypeus with weak cellular microsculpture and angles on front margin of clypeus less pronounced. The eyes are markedly larger, but this is most easily observed in females where the character is not obscured by the variably enlarged temples of the males. Allen (1949) also reports that the elytra are a noticeably brighter yellow in the field.

Habitat: Dung.

Distribution: Rare and scattered in Wales and England north to Midlands.

5. *Oxytelus sculptus* Gravenhorst, 1806 – Plate 12

Length 4-5 mm. Generally dark brown, elytra usually paler, head and abdomen often darker. Antennal tubercles yellow. Legs yellow. Antennae with segments 1 to 3 pale and shiny with scattered setae, segments 4 to 11 darker with close, short pubescence. Easily recognised within *Oxytelus* and *Anotylus* by triangular shape of head which is much narrower than pronotum. Head sparsely punctured with striations ill-defined. Clypeus matt due to strong cellular microsculpture, vertex shiny with strong, round punctures of variable diameter that are not confluent. Front margin of clypeus more convex than other species in the genus. Side margins of pronotum feebly crenulate toward hind angles. Pronotum sparsely punctured and shining with feeble striations confined to sides. Elytra sparsely punctured with fine, irregular striations and short setae. Abdominal tergites clothed with short, but not particularly sparse pubescence. Males with two narrow incisions on the apical margin of sternite VIII (see Fig. 58).

Habitat: Compost heaps and dung heaps.

Distribution: Scattered over Britain and Ireland, but rare or absent in many areas.

Biology: Recorded at light (Lane & Mann, 2006).

6. *PLATYSTETHUS* Mannerheim, 1830

Similar in body form to *Oxytelus*, *Anotylus* and *Aploderus*, but easily distinguished by the central channel and rounded hind angles of the pronotum. In addition, the elytra overlap slightly in repose and the sutural angles are rounded. Abdominal tergites with cellular microsculpture. Apical margin of abdominal tergite VIII weakly concave revealing tergite IX. Males have modifications to abdominal sternite VIII and in several species to sternite VII. Males of some species have two forward-pointing spines on the clypeus. As in *Anotylus* and *Oxytelus*, males also have larger heads with expanded temples making the eyes look smaller. The British and Irish species were reviewed by Hammond (1971), who also provided ecological information and suggested that *Platystethus* species are saprophagous. Worldwide this genus contains 51 species. Seven species in two subgenera have been recorded in the Britain, three of which have been recorded from Ireland.

Key to species of *Platystethus*

1. Elytra, pronotum and head shining with microsculpture absent 2

- Clypeus matt due to strong cellular microsculpture; elytra and pronotum at least with some microsculpture 4

2. Hairs on elytra long and curved, more or less overlapping; punctures on pronotum and sides of head strong and dense tending to become confluent; male abdominal sternite VIII with long spines (see Fig. 66) .... 2. *capito* Heer (p. 41)

- Hairs on elytra short and very sparse, not nearly overlapping; punctures on head and pronotum grouped into clusters, but never confluent; spines on male abdominal sternite VIII short or absent 3

3. Punctures on elytra much smaller and mostly separated by intervals that are considerably greater than their diameters (Fig. 60); male abdominal sternite VIII with short spines (see Fig. 66) .... 5. *nitens* (Sahlberg) (p. 42)

60

- Punctures on elytra larger and grouped into clusters where the intervals are as great as or smaller than their diameters (Fig. 61); male abdominal sternite VIII without spines (see Fig. 66) .... 6. *nodifrons* Mannerheim (p. 42)

61

4. Head lacking furrow inside each eye (Fig. 62); microsculpture on pronotum and elytra limited to longitudinal striations, which are sometimes sparse; males with central tooth on clypeus .... 7. *arenarius* (Fourcroy) (p. 42)

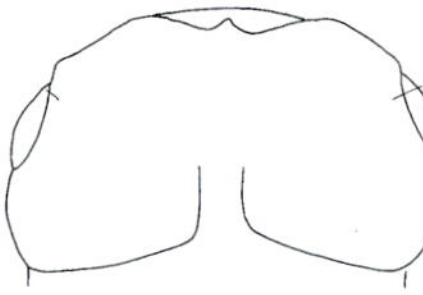

- Head with narrow furrow inside each eye continuing toward back of head (Fig. 63); pronotum and elytra completely covered by cellular microsculpture; males with two long spines on front margin of clypeus 5

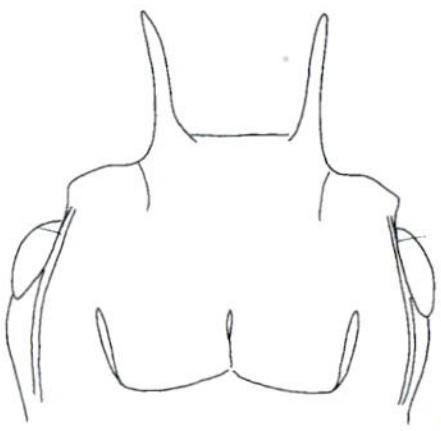

5. Elytra unicolour, usually black; head, pronotum and elytra duller due to stronger microsculpture .... 1. *alutaceus* Thomson (p. 41)

- Elytra black with usually extensive yellow mark; head, pronotum and elytra more shiny due to weaker microsculpture 6

6. Hind margin of male abdominal sternite VII with central, well defined concavity (Fig. 64) .... 3. *cornutus* (Gravenhorst) (p. 41)

- Hind margin of male abdominal sternite VII straight, but with two small tubercles (Fig. 65) .... 4. *degener* Mulsant & Rey (p. 41)

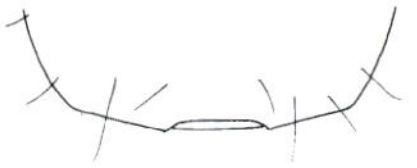

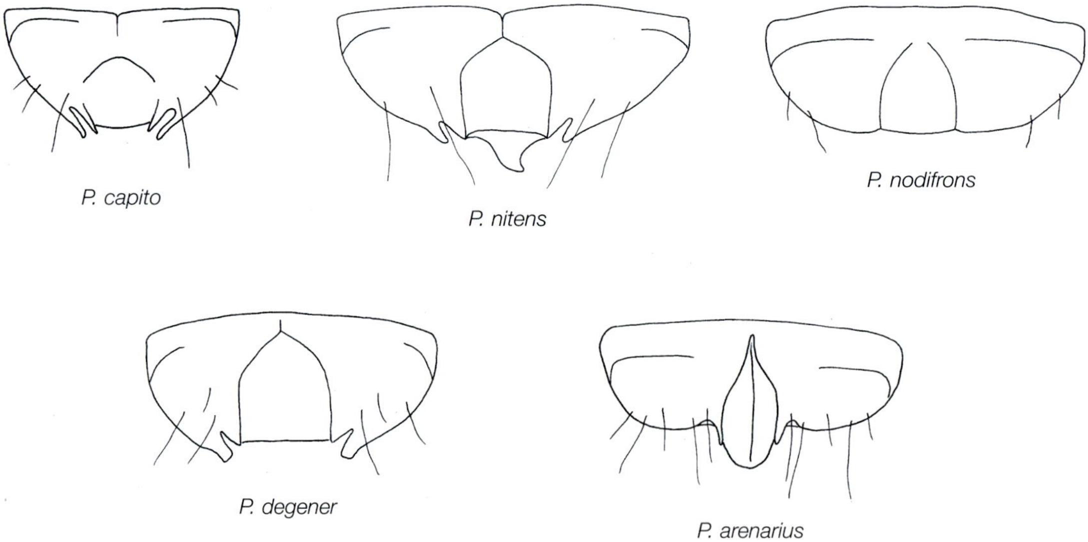

Figure 66. Male abdominal sternites VIII of *Platystethus* species (*P. alutaceus* and *P. cornutus* are similar to *P. degener*).

Subgenus *Craetopycrus* Tottenham, 1939

1. *Platystethus alutaceus* **Thomson, 1861**

Length 3.3-4.8 mm. Body and antennae black, femora dark brown, tibiae brown and tarsi yellow. Males with long spines on front margin of clypeus. Males with modifications to abdominal sternite VII similar to *P. degener*.
Similar species: Larger and duller than *P. cornutus* and never with pale patches on the elytra.
Habitat: Mud in marshes and by ponds.
Distribution: S. England, Wales, S. Scotland.

2. *Platystethus capito* **Heer, 1839**

Length 3 mm. All black except for brown legs and yellow tarsi. Easily recognisable by the longer pubescence and the shape of the pronotum, whose sides are distinctly straightened toward the base. Punctures on head and pronotum irregular but strong and locally confluent. Abdominal tergites relatively shiny by comparison with other species in the genus. Males lack spines on the clypeus.
Habitat: Insolated dry ground. Apparently restricted to chalk and limestone in Britain (Hammond, 1971).
Distribution: S. England; possibly declining.

3. *Platystethus cornutus* **(Gravenhorst, 1802)** – Plate 13

Length 2.2-4.5 mm. Black except for paler legs and elytra which have a yellow mark extending from the inner angle of each elytron toward the shoulder. The extent of this mark is somewhat variable, but is generally diagonal, even if it becomes darker toward the shoulder. Males with long spines on front margin of clypeus.
Habitat: Breeds in exposed mud and sand by ponds and rivers. Large numbers have been found hibernating in leaf litter over 100 metres from a reservoir, the nearest breeding site, suggesting that it may fly to hibernation sites (Lott, 2003).
Distribution: Widespread in lowland areas of Britain, S. Ireland; the most frequently encountered wetland *Platystethus* species.

4. *Platystethus degener* **Mulsant & Rey, 1878**

Length 2.2-4.5 mm. Coloration as *P. cornutus*. Males with tubercles near the apical margin of sternite VII. (The distance between them can vary).
Similar species: Best separated from *P. cornutus* on male sexual characters. The yellow mark on the elytra tends not to extend toward the shoulder and so appears less diagonal, but this difference can be difficult to appreciate, because of infraspecific variability in the extent of the mark.
Habitat: Exposed mud and sand by ponds and rivers, possibly favouring a higher organic content in the substrate than *P. cornutus* (Hammond, 1971).
Distribution: England north to Leicestershire and S. Wales, but predominantly eastern.

5. *Platystethus nitens* (Sahlberg, 1832)

Length 2.5-3 mm. Head, pronotum and abdomen black, elytra often brown. Legs paler. Head with several coarse longitudinal striations inside eye. Mandibles appearing bifid with subapical tooth extending close to apex. Males with two forward-pointing spines (sometimes reduced) on front margin of clypeus.

Habitat: Undisturbed pond margins and marshes with fluctuating water levels. Once found hibernating in numbers under the bark of a beam in a farm outbuilding (leg. A.B. Drane).

Distribution: England north to Yorkshire, S. Wales.

Biology: Taken in light traps (Welch, 1977).

6. *Platystethus nodifrons* Mannerheim, 1830

Length 2.5-4 mm. Coloration as *P. nitens.*

Similar species: More robust and more strongly punctured than *P. nitens.* The pronotum is more transverse. Males lack any forward-pointing spines on the front margin of clypeus.

Habitat: Undisturbed pond margins and marshes with fluctuating water levels.

Distribution: S. England north to Yorkshire, N. Wales, N. Scotland, W. Ireland.

Subgenus *Platystethus* Mannerheim, 1830

7. *Platystethus arenarius* (Fourcroy, 1785) – Plate 14

Length 2.5-4 mm. Black except for brown femora, yellow tibiae and elytra which are brown to yellow, sometimes darkened near scutellum and outer angles. Separated from all other species of *Platystethus* by the lack of a furrow inside the eyes and the non-cellular longitudinal microsculpture on the pronotum and elytra. Pronotum, elytra and abdomen with long lateral setae. Males have a central tooth on the clypeus and an unmodified abdominal sternite VII, while sternite VIII has two short spines inside narrow incisions on the apical margin.

Habitat: Herbivore dung.

Distribution: Widespread in Britain and Ireland.

Biology: Both adults and larvae feed directly on dung and larvae scavenge facultatively on any dead insects that they find (Hinton, 1944). Females exhibit parental care by constructing brood chambers within dung and protecting eggs and first instar larvae from intruders and fungal attack (Hinton, 1944). Taken in flight at sunset (Omer-Cooper & Tottenham, 1934).

7. *APLODERUS* Stephens, 1833

Similar to the larger species of *Oxytelus* or *Anotylus*, but the pattern of depressions on the pronotum is different and, in particular, there is no central furrow. This genus contains 12 species restricted to the Holarctic and Oriental regions. Only one species has been recorded in Britain and Ireland.

1. *Aploderus caelatus* (Gravenhorst, 1802) – Plate 15

Length 4-4.5 mm. Legs yellow. Antennae dark with basal segment red underneath. Head and pronotum black with strong surface sculpture. Pronotum strongly punctured with punctures grouped into clusters. Elytra yellow to dark red, closely punctured. Abdomen with cellular microsculpture on tergites, dark brown with pale hind margins to apical tergites, tergite VIII with straight hind margin. Yellow pubescence generally evident and marginal setae long. Head often enlarged in males, as in species of Oxytelini.
Habitat: In dung or litter.
Distribution: Old records from England, Scotland and SW. Ireland, but there are few recent records away from East Anglia and SE. England and it appears to be declining.
Biology: Taken in flight at sunset (Omer-Cooper & Tottenham, 1934).

8. *BLEDIUS* Leach, 1819

Adaptations to burrowing can be found in several oxyteline genera, but they are most developed in this genus. The prothorax is attached to the hind body on a pedicel that allows the fore body to swivel more freely than other Staphylinidae and this gives *Bledius* species a characteristic look that distinguishes them from all other British staphylinid genera. The front and mid tibiae have a long flat surface fringed by strong spines on both sides, but field observations suggest that they usually use their mandibles for digging (Herman, 1986). The first antennal segment is very long relative to the other segments. Cellular microsculpture is present on the head, pronotum and abdominal tergites. Pubescence is generally well developed on the dorsal surface of the body.

Several species of *Bledius* exhibit striking secondary sexual characters. Males of the subgenera *Bledius* s. str., *Euceratobledius* and *Elbidus* can be armed with a variety of horns on the head and pronotum. The pronotal horn, when present, is long and thin and passes over the top of the head from the front margin of the pronotum. On the head the horns are mounted above each eye. In the subgenus *Elbidus* these are wide and flat and are best described as lamellae. In some species of the subgenus *Hesperophilus*, the males have broad membrane-filled incisions, sometimes with lateral spines on the hind margin of abdominal sternite VII, and these are useful for species identification.

The male genitalia can vary considerably between species groups, but exhibit more subtle interspecific variations within species groups. They contort easily, when they become desiccated because of their poor chitinisation and the fact that key features are twisted in three dimensions. Even when rehydrated, they can remain distorted to some degree. For this reason, the aedeagi of reference specimens should be prepared carefully and mounted in a liquid medium. To conform to the illustrations in this key, they should be positioned ventral surface uppermost. For species identification, they are best examined using fresh specimens or specimens stored in alcohol or dilute acetic acid. Three types of structure can normally be appreciated in properly prepared material (see Fig. 67). The parameres are attached to the ventral surface and the apices are normally twisted round the median lobe toward the dorsal side. In British subgenera, there are two main forms of the parameres. In *Bledius* s. str., *Euceratobledius*, *Elbidus* and *Dicarenus*, they are long and slender. In the other British subgenera, they are usually expanded into two membranes which envelop the median lobe dorsally. The length of the parameres in relation to the median lobe appears to vary within some species. It is possible that the parameres slide up and down the median lobe to some extent during preparation. Two symmetrical lateral valves are visible in the median lobe and, like the parameres, often twisted round the sides. Internal sclerites in the median lobe, here loosely referred to as the internal armature, are sometimes more difficult to see due to poor chitinisation. When visible, the apex of the internal armature is bifid in the *B. pallipes* complex and rounded or simply pointed in superficially similar species in the subgenera *Hesperophilus* and *Astycops*.

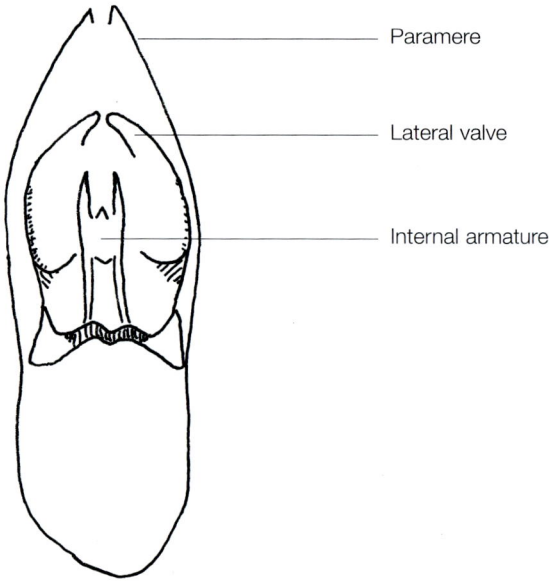

Paramere

Lateral valve

Internal armature

Figure 67. The Aedeagi (ventral aspect) showing the three types of structure normally seen in *Bledius* species.

Across the whole genus, species are liable to exhibit infraspecific variability in several key characters both within individual populations and between geographic areas. Even in the British Isles, it may be difficult to assign individuals to their correct species using the key below. For this reason, it is always preferable to base identifications on a series of specimens. Normally, this should not present problems, because of the gregarious nature of most species. External characters of importance for species identification include secondary sexual characters on the head, pronotum and abdominal sternite VII. The density of punctures on the pronotum and elytra is useful for separating several species, as is the strength of background microsculpture and how this affects the visibility of the punctures on the pronotum. Also important are the length of the elytra relative to the pronotum, the presence or absence of a central furrow on the pronotum and the overall shape of the pronotum, especially the degree of sinuation in the side margin just in front of the hind angle. Subtle differences in this last character are often best appreciated from a latero-dorsal perspective. The degree of sinuation can appear quite different if the orientation of the specimen is even slightly affected by how it is set on the card. The pronotum is often set in a slightly different plane from the elytra because of the flexible articulation of the prothorax and the mesothorax.

Bledius species characteristically burrow into bare sand or clay at water margins, although some species also occur in sand pits well away from permanent water. Most species are found in large colonies. With practice, these colonies can be located by looking for tiny casts of sand at the burrow entrances. Several species are exclusively coastal and some are littoral, their burrows being submerged for short periods at high tide. They feed on algae (Krogerus, 1925; Bro Larson, 1936). There is evidence that some coastal species in Denmark go through at least two generations a year (Bro Larson, 1936). The defensive secretions from abdominal glands have been investigated for several species (Steidle & Dettner, 1995a, 1995b).

The genus includes 453 species distributed worldwide. Herman (1986) abandoned traditional subgenera in his world revision and set up species groups. However, for the British fauna, the old subgenera do not cut across these species groups, so they are retained as a convenient way of grouping species descriptions. The taxonomy of western Palaearctic species in some groups is still in a state of flux. The species boundaries of freshwater species in the British Isles have recently been reviewed (Lott, 2008b) using the male genitalia. 27 species are now recognised as established in Britain, 14 of them also occurring in Ireland, but some coastal species pairs are difficult to separate and are in need of taxonomic revision. In addition, aberrant freshwater specimens or series occasionally turn up and these may represent additional species.

Key to species of *Bledius*

1. Head and pronotum armed with horns or lamellae; [seashore species] 2

- Horns and lamellae absent on head and pronotum 8

2. Fronto-clypeus (rear margin defined by suture between eyes) less transverse; front angles of pronotum produced well forward (Fig. 68); male with long central horn on front of pronotum and horn pointing upwards from head (see Fig. 92) ....
.... 5. *furcatus* (Olivier) (p. 54)

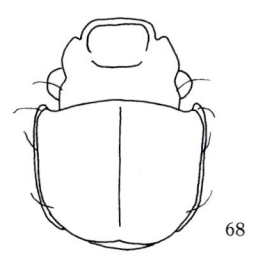

68

- Fronto-clypeus more transverse; front angles of pronotum barely produced forward (Fig. 69); male sexual characters on head and pronotum different 3

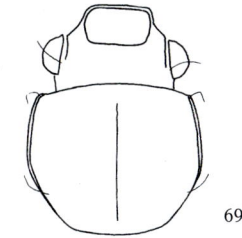

69

3. Pubescence on pronotum longer, semi-erect, but overlapping; pronotum of male lacking horn, but head armed with lamellae which point upwards from the head 4

- Pubescence on pronotum much shorter, semi-erect and not overlapping; pronotum of male with long, central forward-pointing horn and horns on head pointing forwards or reduced to triangular prominences 5

4. Rim on front edge of clypeus missing in centre; male lamellae more quadrangular with a prominent shoulder near the apex and narrower at base above the eye (see Fig. 92) .... 6. *bicornis* Germar (p. 54)

- Rim on front edge of clypeus complete; male lamellae more triangular with a broader base above the eye (see Fig. 92) .... 7. *diota* Schiødte (p. 55)

5. Elytra unicolorous, usually black; length < 4 mm .... 4. *unicornis* (Germar) (p. 54)

- Elytra red with triangular black mark around scutellum sometimes extending to shoulders; length. > 5.5 mm 6

6. Elytra more sparsely punctured and pubescent, punctures on disc separated from each other by more than their own diameter; aedeagus larger (see Fig. 93) .... 2. *spectabilis* Kraatz (p. 53)

- Elytra more densely punctured and pubescent, punctures on disc separated from each other by roughly their own diameter; aedeagus smaller (see Fig. 93) 7

7. Male with horn on head reduced to triangular prominence which merely covers the antennal insertion (see Fig. 92) .... 1. *limicola* Tottenham (p. 53)

- Male with prominent horns on head which point in front of antennal insertion (see Fig. 92) .... 3. *tricornis* (Herbst) (p. 53)

8. Hind margin of tergite VII with dense fringe of hairs (easily visible at 40x magnification though appearing as a membrane at lower magnifications) that is broader in a central emargination (Fig. 70) (females of horned species) 3

- Hind margin of tergite VII with dense fringe of hairs, but central emargination absent (Fig. 71) 9

9. Pronotum with complete central furrow 10

- Pronotum with central furrow absent or confined to front half of pronotum (a complete unpunctured strip may be present) 28

10. Hind angles of pronotum smoothly rounded or very obtuse with side margins in front convex or straightened, but not definitely sinuate (Fig. 72) 11

72

- Hind angles of pronotum sharp or toothed with side margins in front sinuate (Fig. 73) 16

73

11. Pronotum widest in middle and noticeably wider than head, side margins rounded so that they converge towards front and straightened immediately in front of hind angles which are just discernible (Fig. 74); pubescence on elytra sparser with hairs shorter than distance to next puncture 12

74

- Pronotum only slightly wider than head, more or less parallel-sided or even diverging in front of mid-point, hind angles rounded with side margin in front rarely straightened (Fig. 75); pubescence on elytra denser with hairs more or less overlapping 13

75

12. Punctures on pronotum stronger and clearly visible against the background microsculpture, mostly separated from their nearest neighbour by less than their diameter; pronotum quadrate or elongate .... 20. *longulus* Erichson (p. 58)

- Punctures weaker and somewhat lost in the background microsculpture, mostly separated from their nearest neighbour by more than their diameter; pronotum transverse .... 21. *opacus* (Block) (p. 58)

13. Punctures on pronotum shallower and sparser, separated from their nearest neighbour by more than half their diameter; pronotum orange to brown and paler than head; [length 3-4 mm] 14

- Punctures on pronotum stronger and closer, separated from their nearest neighbour by less than half their diameter over most of the pronotum; pronotum and head brown to black, pronotum rarely paler than head 15

14. Abdominal tergites with cellular microsculpture easily visible at 20x magnification .... 9. *praetermissus* Williams (p. 55)

- Abdominal tergites with cellular microsculpture barely discernible at 20x magnification .... 8. *atricapillus* (Germar) (p. 55)

15. Length 4-5 mm; elytra black or bright red and black, pronotum black; male abdominal sternite VII without spines on apical margin, but with central membrane-filled emargination .... 14. *gallicus* (Gravenhorst) (p. 57)

- Length 3-4 mm; elytra brown, pronotum often also brown; male abdominal sternite VII with two apical spines either side of narrow membrane .... 12. *femoralis* (Gyllenhal) (p. 56)

16. First antennal segment darker than apical segments 17

- First antennal segment yellow; antennae unicolorous or darker toward apex 19

17. Labrum bilobed (Fig. 76); elytra all black; length > 4 mm .... 25. *subterraneus* Erichson (p. 59)

76

- Labrum simple (Fig. 77); elytra partly yellow with black area around suture sometimes extending over most of elytra; length < 3.5 mm 18

77

18. Side margin of pronotum in front of hind angle more weakly sinuate (Fig. 78); lateral valves of aedeagus with angled outer margins and truncate apices, normally darkened around mid point and on inner margins near apices (see Fig. 93, but note that these features may be more easily seen by looking at the dorsal aspect) .... 26. *fergussoni* Joy (p. 60)

78

- Sides of pronotum in front of hind angle more strongly sinuate (Fig. 79); lateral valves of aedeagus more rounded and broader without dark markings (see Fig. 93) .... 27. *subniger* Schneider (p. 60)

79

19. Pronotum quadrate **and** antennae entirely yellow 20

- Pronotum transverse **or** antennae darkened toward apex 23

20. Elytra red contrasting strongly with black pronotum and abdomen .... 20. *longulus* Erichson (p. 58)

- Elytra black to brown and more or less unicolorous with pronotum and abdomen 21

21. Elytra densely punctured with distances between punctures less than their diameter; side margin of pronotum in front of hind angle weakly sinuate; aedeagus with lateral valves much shorter and narrower than parameres (see Fig. 93) ... 17. *defensus* Fauvel (p. 57)

- Elytra sparsely punctured with distances between punctures equal to or greater than their diameter; side margin of pronotum in front of hind angle strongly sinuate; aedeagus with lateral valves almost as long as parameres 22

22. Punctures standing out against surface sculpture of pronotum; apex of internal armature of aedeagus with bifid points parallel, proximate and sharp (see Fig. 93); outer margins of lateral valves convex and smoothly rounded, noticeably widest in front of mid-point; length 3.5-4 mm .... 23. *terebrans* (Schiodte) (p. 59)

- Punctures almost lost in surface sculpture of pronotum; apex of internal armature of aedeagus with bifid points diverging and blunter (see Fig. 93); outer margins of lateral valves rounded more unevenly forming characteristic shoulder; length 4-4.5 mm .... 15. *annae* Sharp (p. 57)

23. Elytra red to dark brown, rarely black; aedeagus with more elongate lateral valves that have parallel or sub-parallel outer margins ... 24

- Elytra black, rarely dark brown; lateral valves of aedeagus with more rounded outer margins 25

24. Elytra elongate; pronotum less transverse (Fig. 80) ....
.... 21. *opacus* Block (p. 58)

80

- Elytra quadrate; pronotum more transverse (Fig. 81) ....
.... 16. *arcticus* Sahlberg (p. 57)

81

25. Mid tibiae more slender; tarsi, especially tarsal claws, longer relative to tibiae (Fig. 82) .... 18. *filipes* Sharp (p. 57)

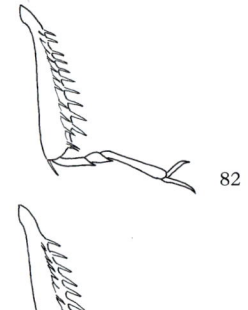

82

- Mid tibiae stouter; tarsi, especially tarsal claws, shorter relative to tibiae (Fig. 83) 26

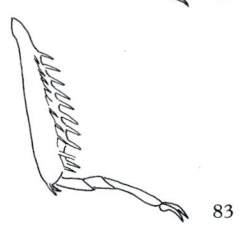

83

26. Elytra less than 1.4x length of pronotum **and** side margin of pronotum strongly sinuate in front of hind angle (Fig. 84); punctures on pronotum often difficult to discern among strong microsculpture; internal armature of aedeagus with diverging, bluntly bifid apex and outer margins of lateral valves with a characteristic shoulder (see Fig. 93) .... 15. *annae* Sharp (p. 57)

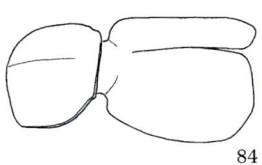

84

- Elytra more than 1.4x length of pronotum **or** side margin of pronotum weakly sinuate in front of hind angle; pronotum more shiny with more distinct puncturation (Figs. 85, 86); internal armature of aedeagus, when visible, with parallel, sharply bifid apex and lateral valves of aedeagus with outer margins more convex and bulging toward apex 27

27. Elytra more than 1.3x length of pronotum; side margin of pronotum distinctly sinuate in front of hind angle (Fig. 85); legs yellow; length 4-4.5 mm .... 22. *pallipes* (Gravenhorst) (p. 59)

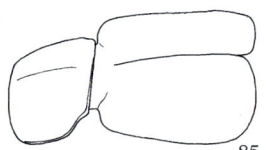

85

- Elytra less than 1.3x length of pronotum; side margin of pronotum weakly sinuate in front of hind angle (Fig. 86); femora often darker than yellow; length 3.5-4 mm .... 19. *fuscipes* Rye (p. 58)

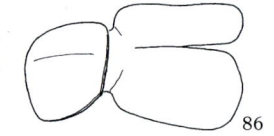

86

28. Hind angles of pronotum sharp with side margins sinuate directly in front (Fig. 87); elytra sparsely punctured with distances between punctures much greater than their diameter; microsculpture on pronotum stronger so that punctures are almost lost in matt background; male with apical margin of sternite VIII simple .... 24. *erraticus* Erichson (p. 59)

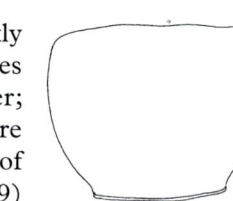

87

- Pronotum with hind angles rounded (Fig. 88); elytra densely punctured with distances between punctures equal to or less than their diameter; microsculpture on pronotum evident, but generally weaker so that punctures are clearly visible on shiny background; male with apical margin of sternite VIII armed with spines or sharply angled ... 29

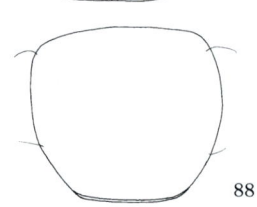

88

29. Elytra longer than pronotum; male with hind angles of abdominal sternite VIII produced into incurving spines (Fig. 89) .... 11. *dissimilis* Erichson (p. 56)

\- Elytra more or less the same length as the pronotum; male with hind angles of abdominal sternite VIII not produced into long incurving spines 30

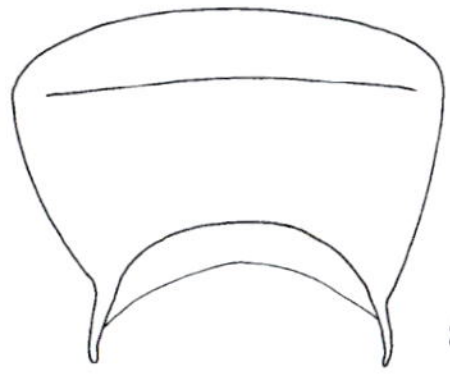

30. Elytra less densely punctured with distances between punctures on shoulder approximately equal to their diameter; male with hind angles of abdominal sternite VIII sharply angled (Fig. 90) .... 13. *occidentalis* Bondroit (p. 56)

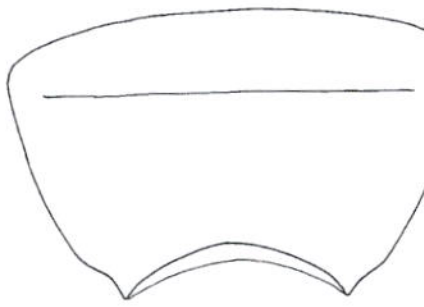

\- Elytra more densely punctured with distances between punctures on shoulder much less than their diameter; male with hind angles of abdominal sternite VIII produced into short but definite spines (Fig. 91) .... 10. *crassicollis* Lacordaire (p. 56)

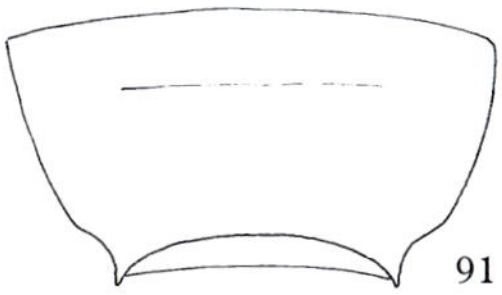

B. limicola

B. spectabilis

B. tricornis

B. unicornis

B. bicornis

B. diota

B. furcatus

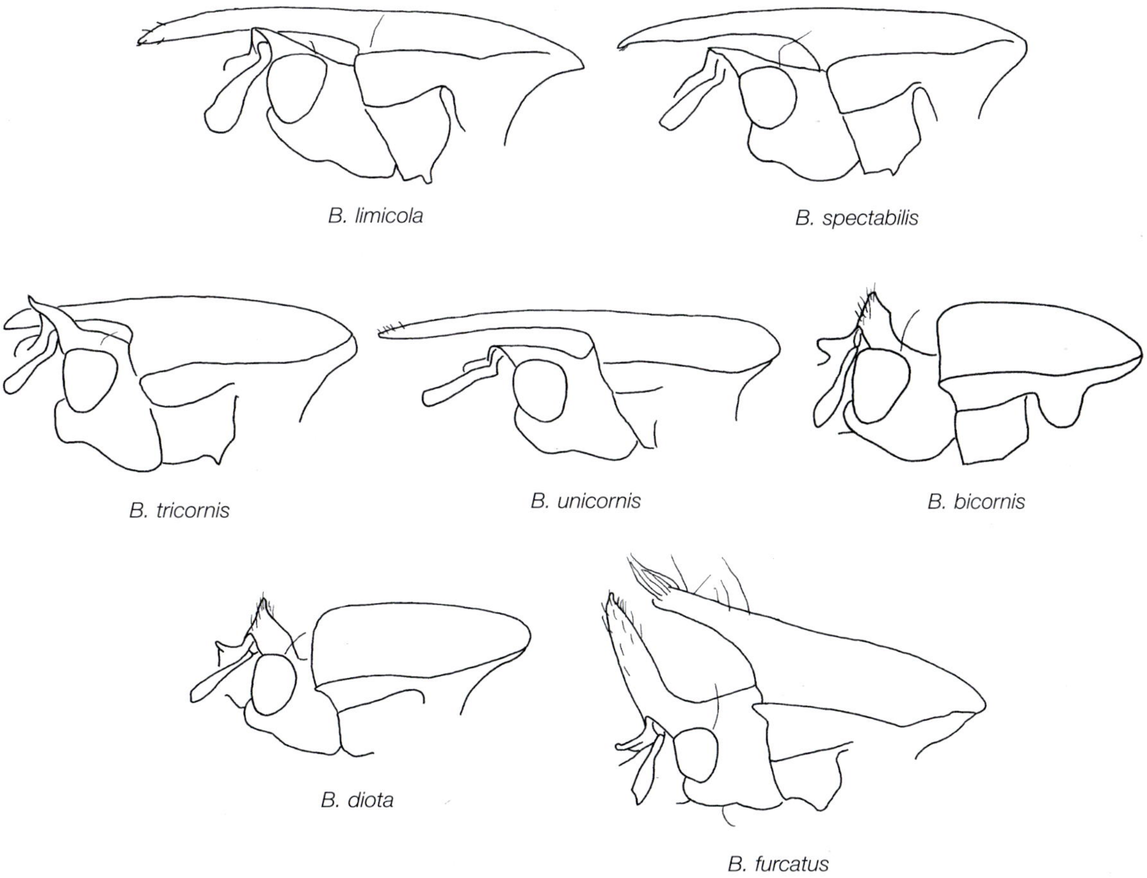

Figure 92. Species of *Bledius* with sexual characters on the male head and pronotum (not drawn to scale).

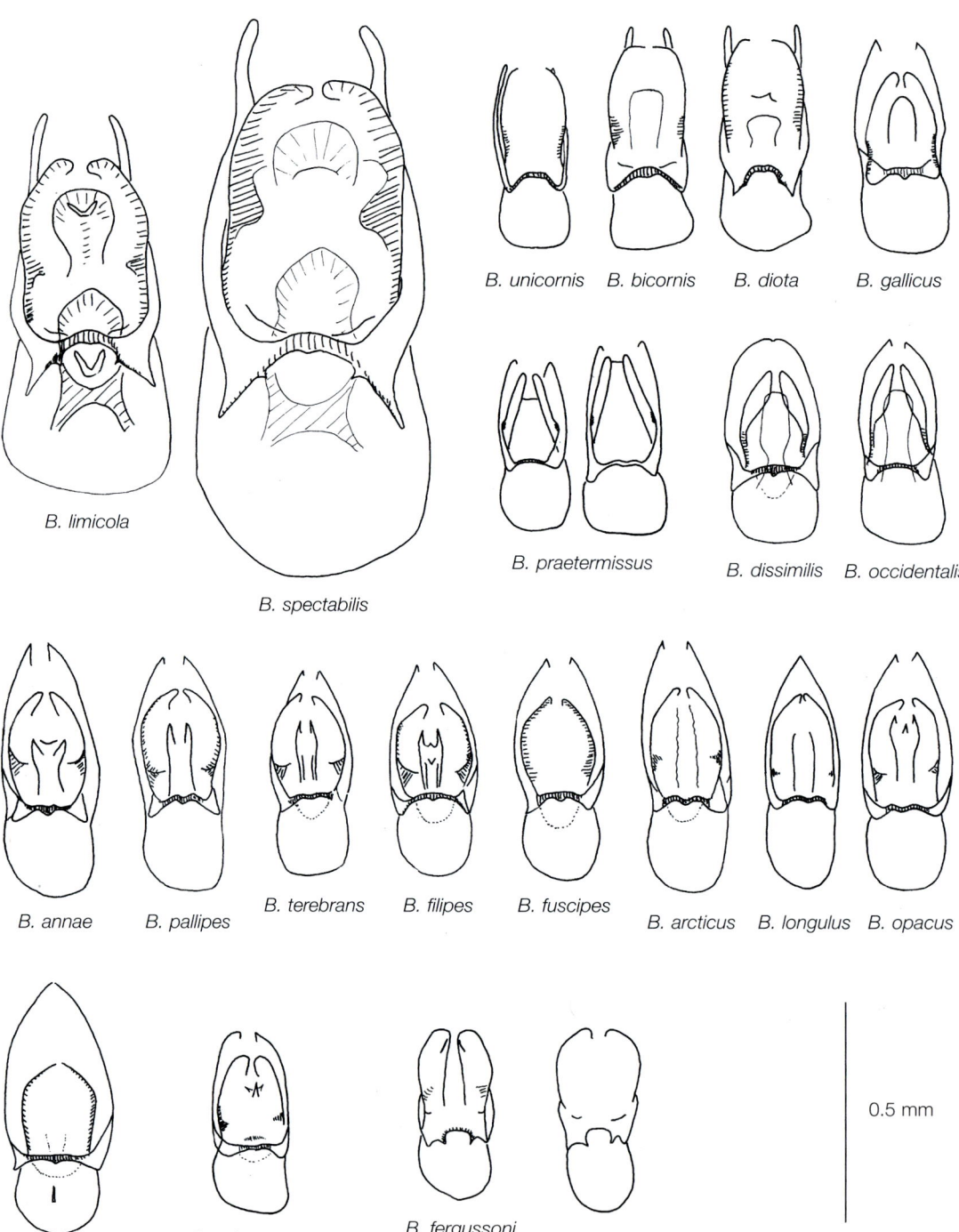

Figure 93. The aedeagi of some species of *Bledius* (ventral aspect).

Subgenus *Bledius* s. str.

This subgenus is equivalent to the *gigantulus* group of Herman (1986) which contains 21 species distributed in the Palaearctic, Oriental and northern Afrotropical regions. Most, if not all, species are restricted to saline habitats. It shares with the subgenera *Elbidus* and *Euceratobledius* a central emargination in the hind margin of abdominal tergite VII. Unlike those subgenera it lacks the pronounced upturned rim on the margin of the clypeus. The males have a straight horn on the front of the pronotum which passes over the top of the head.

1. *Bledius limicola* Tottenham, 1940

Length 5.5-7 mm. Elytra red with black triangular mark around scutellum often extending to shoulders. Head, pronotum and abdomen black. Femora red to dark brown, tibiae and tarsi often paler. Antennae red with basal segment and sometimes other segments darkened. Weak cellular microsculpture present on elytra. Head matt with particularly strong microsculpture. Pubescence yellow and particularly long on abdominal tergites, but absent from head. Male with pronotal horn usually paler at apex, extending to front of head, horn on head reduced to ridged triangular prominence. Aedeagus as in Fig. 93.

Similar species: See *B. spectabilis* and *B. tricornis*.

Habitat: Burrows in bare sand and mud in saltmarshes, estuaries, at the base of sea cliffs and by dune slacks.

Distribution: Around the coast of England, Wales, eastern Scotland and Ireland, where it is one of the most frequently encountered littoral *Bledius* species.

Biology: Occasionally recorded at light, once as far inland as Nottingham (Wright, 1990).

2. *Bledius spectabilis* Kraatz, 1857 – Plates 16-18

Length 7 mm. Aedeagus as in Fig. 93. British populations have been referred to the subspecies *frisius* Lohse by Hammond (2000).

Similar species: Colour and secondary sexual characters as *B. limicola*, except that the dark coloration surrounding the scutellum does not usually extend to the shoulders. Best separated from that species by the larger aedeagus and the sparser puncturation and pubescence on the elytra. The sides of the pronotum are on average straighter or even slightly sinuate behind mid-point.

Habitat: Intertidal sand and mud in saltmarshes and mudflats.

Distribution: Coastal from Anglesey south and east to Norfolk.

Biology: Physiologically and behaviourally adapted to a saline, intertidal environment (Bro Larsen, 1952; Wyatt & Foster, 1988). Both sexes exhibit parental care for the eggs and first instar larvae by protecting them in burrows from predators, anoxia and tidal submersion (Wyatt, 1986, 1989; Wyatt & Foster, 1989).

3. *Bledius tricornis* (Herbst, 1784)

Length 5.5-7 mm. **Similar species:** Colour and aedeagus similar to *B. limicola*. Elytra evenly punctured and without unpunctured areas on the disc, but this condition can also be found in *B. limicola* and many of the key differences described by previous authors and even the secondary sexual characters quoted in this key are not clear cut.

Habitat: Sand and mud in saltmarshes and estuaries, often with *B. limicola*.

Distribution: Coastal from Lancashire south and east to Yorkshire, but much rarer than *B. limicola*.

4. *Bledius unicornis* (Germar, 1825)

Length 3-4 mm. Body black, occasionally with elytra slightly paler. Legs brown with red tarsi. Antennae red to brown. Tubercles above antennal insertions poorly developed. Pronotum quadrate to slightly elongate with rounded hind angles, sometimes with side margins somewhat straightened in front of them. Cellular microsculpture strong. Pubescence pale yellow. Aedeagus as in Fig. 93. Males are hardly likely to be confused with other species in the genus, but females could cause problems if the emarginate apical margin of abdominal tergite VII is overlooked.
Habitat: Intertidal sand and mud in saltmarshes, estuaries and mudflats.
Distribution: Coastal from Lancashire south and east to Norfolk, east coast of Ireland.

Subgenus *Euceratobledius* Znojko, 1929

This subgenus is equivalent to the *furcatus* group of Herman (1986) which includes 10 species distributed in the western Palaearctic, Oriental and eastern Afrotropical regions. Most, if not all species are restricted to saline habitats.

5. *Bledius furcatus* (Olivier, 1811) – Plates 19-20

Length 6-7 mm. Elytra reddish yellow, vaguely darkened around scutellum. Head, pronotum and abdomen black to dark brown. Legs and antennae yellow, femora sometimes darker. An upturned rim extends along the entire margin of the clypeus. Marginal setae on head and pronotum long. Pubescence on pronotum and abdominal tergites much longer than on elytra and absent from head. The male is furnished with horns on both the head and the pronotum, which are reduced or absent in the female. Thick tufts of yellow pubescence adorn the tip of the horn on the pronotum.
Similar species: Female superficially resembles *B. bicornis* but easily distinguished by upturned rim along front margin of clypeus and by the fact that the hairs toward hind margins of elytra are directed inwards.
Habitat: Estuaries, saltmarshes and coastal mudflats.
Distribution: Bred in Norfolk up to 1909, after which time it apparently became extinct in Britain (Hammond in Shirt, 1987). Vagrants, probably from mainland Europe, turn up occasionally in light traps in SE. England (Chuter, 2000).

Subgenus *Elbidus* Mulsant & Rey, 1878

This subgenus is equivalent to the *kochi* group of Herman (1986) which includes 32 species distributed in the Palaearctic and Afrotropical regions. Most, if not all species are restricted to saline habitats. Males have upward pointing lamellae above the eyes. In British species, hairs point backwards over the whole of the elytra.

6. *Bledius bicornis* Germar, 1822 – Plate 21

Length 4.5-5.5 mm. Elytra dark brown with pale sides. Head, pronotum and abdomen brown to dark brown with little contrast between pronotum and elytra. Legs yellow. Antennae yellow, darkened toward apex. Margins of clypeus with upturned rim at sides, but not in front. Pubescence relatively long and pale yellow, absent from head. Aedeagus as in Fig. 93. British populations have been referred to the subspecies *jutlandensis* Herman by Hammond (2000).
Habitat: In saltmarshes and sandy places by the coast.
Distribution: Dorset to Lincolnshire. Old record from Cheshire.
Biology: Recorded at light some distance from the sea (Allen, 1974).

7. *Bledius diota* **Schiødte, 1866**

Length 4-5 mm. Elytra yellow-brown, darker on disc and by suture. Head, pronotum and abdomen black to dark brown with elytra noticeably paler. Legs and antennae yellow. Upturned rim continued along front margin of clypeus. Pubescence similar to *B. bicornis*. Punctures on elytra sparser. Aedeagus as in Fig. 93.
Habitat: In sandy places by the coast.
Distribution: Sussex to Lincolnshire, Somerset.
Biology: Recorded at light several kilometres from the sea (Hodge, 1979).

Subgenus *Hesperophilus* Curtis, 1829

This subgenus is equivalent to two of Herman's (1986) species groups. The *semiferrugineus* group includes 57 species, including species 8 to 14 below, and has a distribution centred in the Holarctic region extending south to southern Africa and South America. The *annularis* group includes 73 species distributed throughout the Holarctic region including species 15 to 24 below. Most species have freshwater habitats.

8. *Bledius atricapillus* **(Germar, 1825)**

Length 3-3.5 mm. Relatively narrow-bodied for *Bledius* species. Head black to dark brown. Pronotum brown. Elytra yellow, vaguely darkened around the scutellum and along the suture. Abdominal tergites brown to dark brown. Legs yellow. Antennae yellow with apical segments sometimes darker. Pubescence short on elytra.
Similar species: Apart from the key characters, usually somewhat smaller and more delicate than *B. praetermissus*. Many of the characters quoted in the literature to separate the two species are not always reliable. *B. atricapillus* is normally paler and has a smaller pronotum with more rounded sides. The punctures on the elytra are finer and sparser. However, populations with intermediate characters occur. Illustrations of the aedeagi by Lohse (1982) and Vorst (2003) suggest that in *B. atricapillus* the outer margins of the lateral valves are more strongly elbowed just before mid point, but this character is not constant in *B. praetermissus* (see Fig. 93). Furthermore, slight differences in the orientation of the aedeagus can make it appear much more strongly elbowed. Distortion caused by desiccation will have the same effect.
Habitat: Uncertain. Its range extends further inland in Central Europe than *B. praetermissus* (Lohse, 1982; Vorst, 2003).
Distribution: Unclear, but known from area around the mouth of the Thames. Apparently much rarer than *B. praetermissus*.

9. *Bledius praetermissus* **Williams, 1929**

Length 3-4 mm. Relatively narrow-bodied for *Bledius* species. Head black to dark brown. Pronotum dark brown, sometimes paler. Elytra yellow to yellow-brown, sometimes vaguely darkened around the scutellum and along the suture. Abdominal tergites brown to black. Legs yellow. Antennae yellow to brown with first two segments often paler. Pubescence rather short on elytra. Aedeagus as in Fig. 93.
Similar species: See *B. atricapillus*.
Habitat: By seepages on cliffs and sandy riverbanks close to the sea.
Distribution: English coast north to Solway and Yorkshire, Galloway, Isle of Man, E. Ireland. Often occurs in large colonies.

10. *Bledius crassicollis* Lacordaire, 1835

Length 3-3.5 mm. Elytra reddish, dark basal markings either absent or not extending along suture. Head and pronotum black. Abdominal tergites black with basal segments reddish. Legs and antennae yellow. Pubescence long and relatively dense. Head with a narrow shiny strip behind the suture separating the vertex and fronto-clypeus. Pronotum closely and regularly punctured leaving a well defined shiny central stripe through the mid line. Microsculpture on pronotum confined to area behind front margin.
Similar species: Generally smaller than *B. dissimilis* and *B. occidentalis*. Apart from the key characters, distinguished from both species by the closer, more regular punctures on the pronotum.
Habitat: Recorded from coastal freshwater seepages in Britain, but known mainly from gravel pits in C. Europe and from riverbanks in S. Europe.
Distribution: Only known from Kent and the Isle of Wight. Very rare.

11. *Bledius dissimilis* Erichson, 1840

Length 3-4 mm. Elytra yellow or red with diffuse dark mark of varying extent, often extending from base along suture, sometimes covering most of elytra. Head and pronotum generally black. Abdominal tergites black to reddish. Legs and antennae yellow. Pubescence long and relatively dense. Head with a narrow shiny strip behind the suture separating the vertex and fronto-clypeus. Density of punctures on pronotum and elytra somewhat variable. Pronotum rather irregularly punctured leaving small patches unpunctured and an ill defined shiny central stripe through the mid line. Microsculpture on pronotum confined to area behind front margin. Aedeagus as in Fig. 93.
Similar species: The longer and generally darker elytra distinguish this species from *B. crassicollis* and *B. occidentalis*.
Habitat: Recorded from seepages on clay cliffs and from gravel pits and quarries.
Distribution: SE. England to Yorkshire. The frequency of records has increased in recent years.
Biology: Has been recorded in numbers at light (Lane & Mann, 2006).

12. *Bledius femoralis* (Gyllenhal, 1827)

Length 3-4 mm. Elytra uniformly brown. Pronotum brown to black. Head and abdominal tergites black. Legs yellow. Antennae variable in colour.
Similar species: *B. femoralis* has a central furrow on the pronotum like *B. gallicus*, but is smaller and paler with a narrower pronotum and more sparsely punctured elytra.
Habitat: In damp sandy areas.
Distribution: S. England, S. Wales and Yorkshire.

13. *Bledius occidentalis* Bondroit, 1907

Length 3-3.5 mm. Elytra reddish with dark basal markings either absent or not extending along suture. Head, pronotum and abdomen black. Legs and antennae yellow. Pubescence long and relatively dense. Head completely matt. Pronotum rather irregularly punctured leaving small patches unpunctured and an ill defined shiny central stripe through the mid line. Microsculpture on pronotum confined to area behind front margin. Aedeagus as in Fig. 93.
Similar species: See *B. dissimilis* and *B. crassicollis*
Habitat: Sand dunes and undercliffs.
Distribution: Mostly coastal between Devon and Yorkshire, also Cambridgeshire. Recently recorded inland in Ireland on bare peat (Regan & Anderson, 2004).

14. *Bledius gallicus* (Gravenhorst, 1806) – Plate 22

Length 4-5 mm. Elytra with two colour forms: the type form is all black and var. *laetior* Mulsant & Rey is red with triangular black mark around scutellum. They may be separate species. Head, pronotum and abdominal tergites black in both colour forms. Legs yellow to red. Antennae red to brown. Pubescence relatively long. Aedeagus as in Fig. 93.

Similar species: The rounded shape of the pronotum and the rounded apex to the internal armature of the aedeagus will distinguish this species from superficially similar freshwater species.

Habitat: In sand by streams and rivers, sandpits, small patches of bare ground in marshes.

Distribution: Var. *laetior* is widely distributed in the British Isles, but the type form is confined to southern England, Wales and the Isle of Man. They rarely occur together.

15. *Bledius annae* Sharp, 1911

Length 4-4.5 mm. Head, pronotum, elytra and abdomen black to dark brown. Legs and antennae all red. Aedeagus as in Fig. 93.

Similar species: Distinguished from *B. pallipes* by shorter and narrower elytra, narrower pronotum, which has less distinct punctures against the strong surface microsculpture, and the internal armature and lateral valves of the aedeagus.

Habitat: In sand by larger rivers.

Distribution: Speyside, S. Scotland, Northumberland and Worcestershire.

16. *Bledius arcticus* Sahlberg, 1890

Length 4.5-5 mm. Elytra yellow, red or dark brown, often vaguely darkened at base and along suture (individual populations usually contain a mixture of colour forms). Head, pronotum and abdomen black. Legs yellow. Antennae yellow, darkened toward the apex. Aedeagus as in Fig. 93.

Similar species: Distinguished from *B. pallipes* by colour pattern and shorter elytra. Distinguished from *B. opacus* by the shape of the pronotum and the shorter elytra.

Habitat: Sandy areas on river shingle banks.

Distribution: Scottish Highlands and Hebrides.

17. *Bledius defensus* Fauvel, 1872

Length 3.5-4 mm. Elytra brown to dark brown. Head, pronotum and abdomen black to dark brown. Legs and antennae yellow-red. Aedeagus distinctive with short lateral valves compared with the parameres. Aedeagus highly characteristic (see Fig. 93).

Similar species: Distinguished from *B. pallipes* by the form of the aedeagus, smaller overall size and narrower pronotum which is more closely punctured so that the punctures often form rows near the central channel. Usually paler and sides of pronotum more weakly sinuate in front of hind angles than *B. annae*.

Habitat: Typically burrows into shaded, vertical riverbanks in Britain.

Distribution: N. England and N. Midlands.

18. *Bledius filipes* Sharp, 1911

Length 3.5-4 mm. Head, pronotum, elytra and abdomen black. Legs yellow-red. Antennae dark, basal segments yellow. Pronotum quadrate or only slightly transverse, such that the front margin is not much wider than the head. Side margin of pronotum in front of hind angle definitely sinuate. Elytra more than 1.3x longer than pronotum. Aedeagus as in Fig. 93.

Similar species: Distinguished from *B. pallipes* by smaller size, narrower pronotum, more slender legs and longer tarsi. Aedeagus similar to *B. pallipes*, but lateral valves more parallel sided.

Habitat: Confined to soft rock coastal cliffs in Britain, but associated with riverbanks in mainland Europe.

Distribution: Norfolk.

19. *Bledius fuscipes* Rye, 1865

Length 3.5-4 mm. Elytra black, occasionally dark brown. Head, pronotum and abdomen black. Legs red to brown. Antennae dark, basal segment yellow. Aedeagus as in Fig. 93.

Similar species: Usually distinguished from *B. pallipes* by smaller size, shorter elytra and the shape of pronotum. Aedeagus similar to *B. pallipes*, except that the internal armature tends to be weakly sclerotised.

Habitat: Sandy areas including sandpits.

Distribution: Mainly coastal in Britain and W. Ireland; more frequently recorded on western coasts.

20. *Bledius longulus* Erichson, 1839

Length 3-4 mm. Elytra red, vaguely darkened around scutellum. Head, pronotum and abdominal tergites black. Legs yellow. Antennae yellow, sometimes slightly darkened toward apex. Aedeagus as in Fig. 93.

Similar species: Smaller and more slender than *B. opacus*. In particular the pronotum is narrower with stronger punctures that are more easily visible against a more shiny background and with more strongly sinuate side margins in front of hind angle.

Habitat: Sandy riverbanks and quarries, sometimes in association with *B. opacus*.

Distribution: Widely distributed in Britain and N. Ireland, though absent or rare in some areas.

21. *Bledius opacus* (Block, 1799) – Plate 23

Length 3.5-4.5 mm. Elytra yellow to red, darkened around scutellum and often along suture, rarely extending onto disc of elytra. Head, pronotum and abdominal tergites black. Legs yellow. Antennae yellow, often slightly darkened toward apex. Pronotum usually with rounded hind angles, but occasionally straightened or even slightly sinuate in front of obtuse hind angles. Aedeagus as in Fig. 93.

Similar species: The rounded hind angles of the pronotum will usually distinguish this species from other members of the *B. pallipes* group. The odd specimens with more marked angles usually occur in colonies with more typical specimens. Distinguished from *B. gallicus* var. *laetior* by the shape of the pronotum and its less distinct punctures. See also under *B. longulus*.

Habitat: Sandy riverbanks and quarries, sometimes in association with *B. longulus*.

Distribution: Widely distributed in Britain and Ireland, but with a more southern bias than *B. longulus*. One of the more frequently encountered *Bledius* species in artificial habitats.

22. *Bledius pallipes* (Gravenhorst, 1806) – Plate 24

Length 4-4.5 mm. Head, pronotum, elytra and abdomen black. Legs yellow to red. Antennae yellow, normally, though not always, darkened toward the apex. Aedeagus with sharply acute, parallel bifid points to the internal armature and bulging, convex outer margins to the lateral valves, which are widest in their apical half. A rather variable species, against which all new specimens from freshwater margins should be compared. Aedeagus as in Fig. 93.
Habitat: Banks of large rivers.
Distribution: Widely distributed in the British Isles except for NW. Scotland and the most commonly encountered freshwater *Bledius* species in lowland areas.

23. *Bledius terebrans* (Schiødte, 1866)

Length 3.5-4 mm. Elytra black, rarely dark brown. Head, pronotum and abdomen black. Legs and antennae all red.
Similar species: Aedeagus similar to *B. pallipes* (see Fig. 93) from which it can be distinguished by its smaller size, shorter elytra and the narrower pronotum which has characteristically shiny punctures standing out from the microsculpture.
Habitat: In sandy stream banks.
Distribution: Rare, but scattered throughout Scotland and N. England. Old records also from Surrey and Sussex.

24. *Bledius erraticus* Erichson, 1839

Length 3-4 mm. Relatively narrow-bodied and quite distinct within the subgenus. Elytra yellow-brown. Head, pronotum and abdomen reddish brown. Legs yellow. Antennae yellow, sometimes darker toward apex. Pronotum with side margins sinuate in front of well marked hind angles. Central furrow on pronotum absent. Punctures on elytra relatively sparse.
Habitat: In sand and shingle by streams, rivers and lakes.
Distribution: Widely scattered in N. England, Scotland and N. Ireland. Rarely recorded.

Subgenus *Astycops* Thomson, 1859

This subgenus is equivalent to the *albonotatus* group of Herman (1986) which includes 15 species distributed in the Holarctic region. It is characterised by the bilobed labrum. All but one species are associated with freshwater.

25. *Bledius subterraneus* Erichson, 1839 – Plate 25

Length 4-4.5 mm. Elytra, head, pronotum and abdomen black. Legs yellow, occasionally brown. Antennae yellow to brown with basal segment darkened. Aedeagus with single pointed apex to internal armature. Aedeagus as in Fig. 93.
Similar species: Pronotum transverse and elytra long as in *B. pallipes*, but distinguished by the darker basal segment of the antennae, the bilobed labrum and the form of the aedeagus.
Habitat: In sand by rivers and streams.
Distribution: Widely distributed and often abundant in the hilly parts of Britain and N. Ireland, scarcer in lowland areas.

Subgenus *Dicarenus* Gistel, 1834

This subgenus is equivalent to the *basalis* group of Herman (1986) which includes 14 species distributed in the western Palaearctic and Nearctic regions. Most species have saline habitats. The pronotum in British species is more transverse and rectangular than in other subgenera and similar in width to the elytra making the whole insect look somewhat parallel-sided. The antennae are short with a relatively well-marked club.

26. *Bledius fergussoni* Joy, 1912 – Plate 26

Length 2.5-3.5 mm. Elytra pale yellow with a dark mark at base and along suture sometimes extending over most of the elytra. Head, pronotum and abdomen black, relatively shiny. Femora dark brown. Tibiae usually yellow. Antennae yellow to brown with darkened basal segment. Elytra rather finely punctured. Aedeagus as in Fig. 93.
Habitat: In damp sandy places by the coast.
Distribution: Widely distributed in Britain and Ireland around the coast, especially western coasts.

27. *Bledius subniger* Schneider, 1900

Length 3 mm. Aedeagus as in Fig. 93.
Similar species: Coloration as *B. fergussoni*, except that on average the darker area on the elytra is more extensive. Most reliably separated on male genitalia, though the difference in sinuation of the side margin works for most specimens. Punctures on pronotum generally stronger and more evident.
Habitat: Damp sandy places by the coast, saltmarsh.
Distribution: Coasts of Britain and Ireland, but scarcer in E. England.

9. *CARPELIMUS* Leach, 1819

A combination of short legs and antennae, stocky build and small size make this genus easy to recognise in the field. The centre of the pronotum typically has two longitudinal depressions, which in some species can be broken into four round depressions, or in others are so shallow as to be barely visible. In addition, some of the larger species have a lateral depression on the edge of the pronotum either confined to a small area toward the base or stretching forward to the front angles. In many species there is a longitudinal, shallow crease in the front half of the elytra near to the suture. The penultimate segment of the maxillary palps is greatly widened, while the terminal segment is small and needle-like (see Fig. 95). Abdominal tergite VII has a weakly concave hind margin, while the hind margin of the tergite VIII is weakly concave to straight.

There is some variability within species of key characters such as the proportions of antennal segments and the size of eyes and pronotum. Use of a reliable reference collection to aid identification is advised. The male genitalia contain internal sclerites which are of great value in species identification. They are easily visible and rarely need clearing. See Fig. 94 for a schematic diagram showing the internal sclerites visible in the aedeagi of species in the subgenus *Paratrogophloeus*. For reference specimens, the aedeagi are best mounted in a medium such as DMHF. To conform to the illustrations in this key, they should be positioned dorsal surface uppermost. The internal sclerites are three-dimensional structures, so when they are everted, they may appear somewhat different to the illustrations. When partially everted, their position within the median lobe may change. There are no obvious external differences between the sexes, except in two species (*C. incongruus* and *C. zealandicus*).

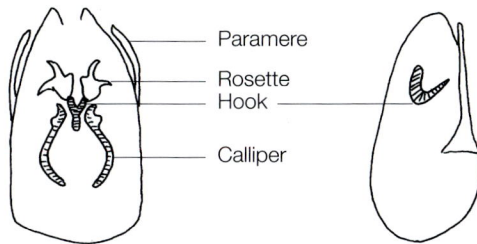

Fig. 94. Internal sclerites visible in the aedeagi of species in the subgenus *Paratrogophloeus*.

Gildenkov (2001a) has reviewed the biology of *Carpelimus* species. They are adapted for tunnelling into soft sediments by water margins and in marshes. Species of the subgenus *Troginus* are more active burrowers than other subgenera that take advantage of existing fissures in sand or finer sediments containing coarse organic matter. They require damp substrates, but avoid waterlogged conditions. Some species are associated with accumulations of litter, but they are rarely found on pure peat. Their habitats vary from wet woodland to sandy riverbanks. Several species are restricted to coastal saline environments. Most, if not all *Carpelimus* species typically feed on semi-decomposed plant and animal matter. Some species have been reported feeding on the green parts of higher plants. Accumulations of organic matter in their habitat often occur unevenly and ephemerally, so *Carpelimus* population densities often fluctuate wildly. In Russia densities of 500 to 800 individuals per square metre have been recorded. Dispersal takes place during the first half of the night, when they can be caught in light traps. Females are more likely to disperse than males.

This genus contains 413 described species of which 19 are currently recorded as established breeding species in Britain, 11 of these also occurring in Ireland. In addition, *C. despectus* (Baudi) has been listed as a British species associated with coastal mud or sand flats in England (Hammond, 2000). Its inclusion was based on a single museum specimen collected at Gretna in 1934 (P.M. Hammond, pers. comm.). Its true British status remains unclear and I have not had an opportunity to see British specimens, so it is not included in the key. A single male *C. alutaceus* (Fauvel) was found in 1990 on the banks of the River Soar near Loughborough in Leicestershire, but there is no evidence that it has an established breeding population in Britain. Both species are listed by Gildenkov (2001a) as halophilic. Gildenkov (2001a) has revised the Palaearctic species.

Key to species of *Carpelimus*

1. Pronotum very weakly arched, with rounded or almost angled sides and greatest width in front of middle, comfortably wider than head (see Figs 95-100); antennal segment 5 elongate, segment 10 quadrate or elongate; [body length 2.5-4 mm] 2

- Pronotum flat or arched, shape variable, but if wider than head then only slightly (see Figs 101-111); antennal segment 5 usually quadrate or transverse, if elongate then segment 10 transverse; [body length 1.5-3 mm] 8

2. Temples very short, when seen from above less than half the diameter of the eyes which project out from the side of the head much further than the temples (see Figs 95, 96) 3

- Temples as long as or longer than half the diameter of the eyes when seen from above, very convex and projecting laterally further than or only slightly less than the eyes (see Figs 97-100). (Five species whose lateral pronotal margins have a well-defined strip of coarse, contiguous punctures contrasting with the distinct punctures on the central area and whose aedeagi contain characteristic, visible internal sclerites as labelled in Fig. 94) 4

3. Legs and antennae black or at least darkened; depressions present in centre and side margin of pronotum (Fig. 95) ....
.... 3. *obesus* (Kiesenwetter) (p. 68)

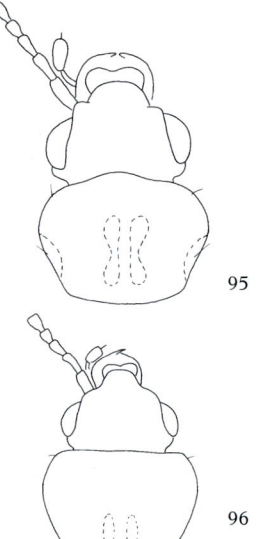

95

- Legs yellow; antennae yellow becoming slightly darker toward apex; side margin of pronotum flat, depressions confined to base (Fig. 96) .... 1. *fuliginosus* (Gravenhorst) (p. 67)

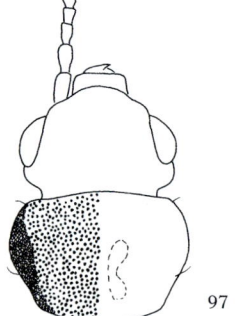

96

4. Eyes and temples flatter, angle between hind margin of eye and temple very obtuse when seen from above; pronotum generally narrower (Fig. 97); antennal segment 5 only slightly elongate; aedeagus with visible rosettes but both callipers and central hook absent (see Fig. 112) .... 17. *impressus* (Boisduval & Lacordaire) (p. 72)

- Eyes and temples more protruding so that angle between hind margin of eyes and temples is around 120 degrees when seen from above (see Figs 98-100); antennal segment 5 more elongate; either callipers or central hook visible within aedeagus 5

97

5. Coarser punctures on pronotal margin extending a good third of the way to the centre (Fig. 98); aedeagus with rosettes, but no central hook. (Two species only separable on the male genitalia) 6

- Coarser punctures on pronotal margin extending only a quarter of the way to the centre (see Figs 99, 100); aedeagus with central hook but no rosettes (see Fig. 112) 7

98

6. Callipers in aedeagus thinner, generally brown; rosettes with spike extending toward apex (see Fig. 112) ....
.... 7. *erichsoni* (Sharp) (p. 69)

- Callipers in aedeagus thicker, generally black; rosettes without spike extending toward apex (see Fig. 112) ....
.... 6. *bilineatus* Stephens (p. 69)

7. Pronotum generally narrower and slightly more arched (Fig. 99); callipers in aedeagus reduced (see Fig. 112) ....
.... 8. *rivularis* (Motschulsky) (p. 69)

99

- Pronotum generally broader and slightly flatter (Fig. 100); callipers clearly visible in aedeagus (see Fig. 112) ....
.... 9. *similis* Smetana (p. 70)

100

8. Punctures on pronotum confused with coarse surface sculpture so that intervals are indiscernible or matt 9

- Pronotum with distinct punctures separated from each other by shiny, if sometimes narrow intervals 14

9. Diameter of eyes less than 1.5x longer than length of temples when seen from above (see Figs 101, 102) 10

- Diameter of eyes 2x or more greater than length of temples when seen from above (see Figs 103, 104, 109, 111) 11

10. Temples generously rounded, projecting laterally well beyond eyes and forming a sharp angle with the neck (Fig. 101); length generally longer, 2-3 mm .... 5. *elongatulus* (Erichson) (p. 68)

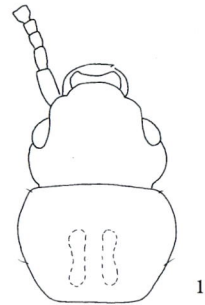

101

- Temples flatter, hardly projecting beyond eyes and not well differentiated from neck (Fig. 102); length generally shorter, 1.7-2.25 mm .... 11. *schneideri* (Ganglbauer) (p. 70)

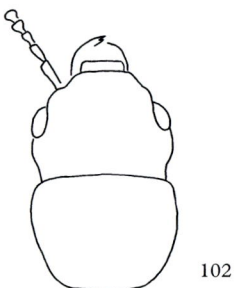

102

11. Legs and antennae predominantly black (tarsi, knees and the apices of the tibiae paler); elytra strongly punctured, but shiny between the punctures and contrasting with matt pronotum; temples strongly rounded and protruding beyond width of eyes (Fig. 103) .... 14. *foveolatus* (Sahlberg) (p. 71)

- Colour of legs and antennae variable, but always paler and predominantly red on front tibiae; elytra more finely and densely punctured, less shiny in comparison with pronotum; temples often rounded, but not protruding beyond width of eyes (see Figs 104, 109, 111) 12

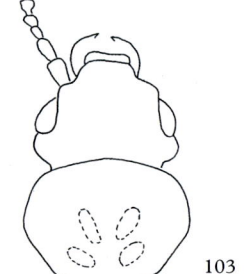

103

12. Elytra strongly and closely punctured so that surface is quite rough, flat near suture; pronotum less transverse, depressions absent or barely discernible (Fig. 104) ....
.... 16. *halophilus* (Kiesenwetter) (p. 72)

- Elytra either finely punctured or less closely punctured with longitudinal crease in front half near suture; pronotum more transverse, depressions sometimes weak, but always evident (see Figs 109, 111) 13

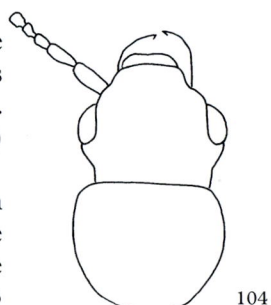

104

13. Abdominal tergites with cellular microsculpture; temples longer in proportion to eyes (see Fig. 111); elytra generally black; aedeagus as in Fig. 112 .... 18. *manchuricus* (Bernhauer) (p. 72)

- Abdominal tergites finely and densely punctured; temples shorter in proportion to eyes (see Fig. 109); elytra brown; aedeagus as in Fig. 112 .... 4. *pusillus* (Gravenhorst) (p. 68)

14. Pronotum highly arched and almost conical in form, sinuate towards base (Fig. 105); third antennal segment markedly thinner than second 15

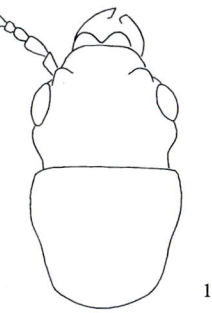

105

- Pronotum weakly arched, rounded at sides or at most, straightened towards base (see Figs 97, 106-111); third antennal segment only slightly thinner than second 16

15. Punctures on inner half of elytra finer, their diameters usually smaller than the distance to the nearest puncture; internal sclerites in aedeagus broader and not obviously hooked at base (see Fig. 112) .... 10. *incongruus* Steel (p. 70)

- Punctures on inner half of elytra stronger, their diameters usually greater than the distance to the nearest puncture; internal sclerites in aedeagus narrower and more obviously hooked at base (see Fig. 112) .... 12. *zealandicus* (Sharp) (p. 71)

16. Diameter of eye less than 1.5x length of temples when seen from above (see Figs 106-108) 17

- Diameter of eye greater than 2x length of temples when seen from above (see Figs 97, 109-111) 19

17. Pronotum widest nearer to middle than front angles (Fig. 106); length >2 mm .... 13. *corticinus* (Gravenhorst) (p. 71)

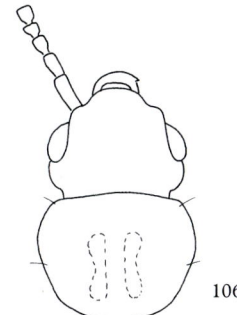

106

- Pronotum widest nearer to front angles than to middle (see Figs 107-108); length < 2 mm 17

18. Diameter of eye roughly equal to length of temple when seen from above; width of head across eyes roughly equal to width of head across temples (Fig. 107) ....
.... 15. *gracilis* (Mannerheim) (p. 71)

107

- Diameter of eye about half the length of the temple when seen from above; width of head clearly widest across temples (Fig. 108) .... 19. *subtilis* (Erichson) (p. 73)

108

19. Abdominal tergites finely and densely punctured; elytra matt brown, also finely and densely punctured; pronotum less transverse (Fig. 109) .... 4. *pusillus* (Gravenhorst) (p. 68)

- Abdominal tergites with cellular microsculpture; elytra generally black with more distinct punctures; pronotum more transverse (see Figs 97, 110-111) 20

109

20. Legs (except for tarsi, knees and apices of tibiae) and antennae black; pronotum strongly punctured, but shiny; temples very short in comparison with eyes (Fig. 110) ....
.... 2. *lindrothi* Palm (p. 68)

- Legs and basal joints of antennae largely red; pronotum duller due to closer puncturation; temples longer in comparison with eyes (see Figs 97, 111) 21

110

21. Pronotum more strongly rounded at sides and more transverse; head more transverse; eyes larger in comparison to temples (see Fig. 97); length > 2.5 mm; aedeagus as in Fig. 112 ....
.... 17. *impressus* (Boisduval & Lacordaire) (p. 72)

- Pronotum less strongly rounded at sides and less transverse; head less transverse; eyes smaller in comparison to temples (Fig. 111); length < 2.5 mm; aedeagus as in Fig. 112 ....
.... 18. *manchuricus* (Bernhauer) (p. 72)

111

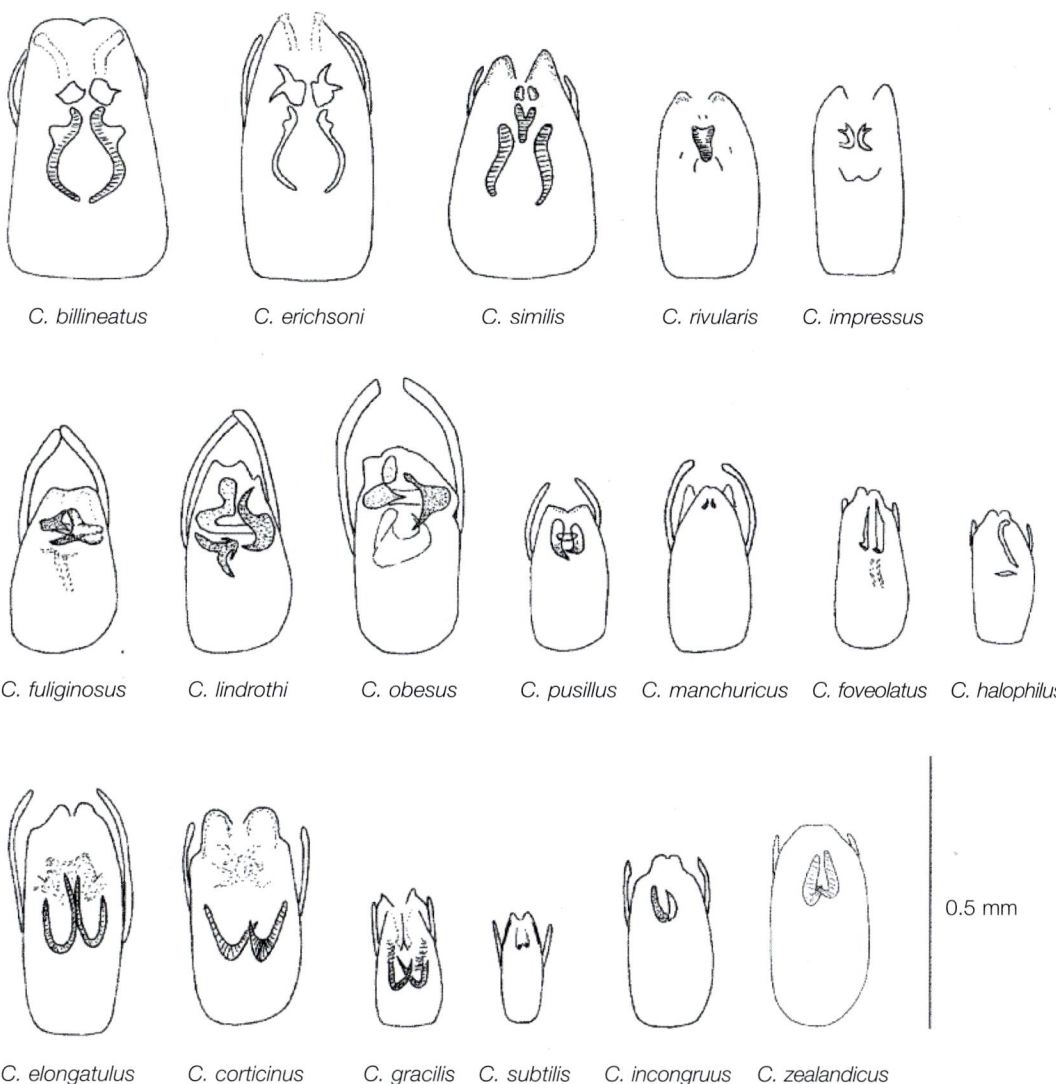

C. billineatus   *C. erichsoni*   *C. similis*   *C. rivularis*   *C. impressus*

C. fuliginosus   *C. lindrothi*   *C. obesus*   *C. pusillus*   *C. manchuricus*   *C. foveolatus*   *C. halophilus*

C. elongatulus   *C. corticinus*   *C. gracilis*   *C. subtilis*   *C. incongruus*   *C. zealandicus*

0.5 mm

Fig. 112. The aedeagi of some species of *Carpelimus* (dorsal aspect).

Subgenus *Carpelimus* Leach, 1819

1. *Carpelimus fuliginosus* (Gravenhorst, 1802)

Length 2.5-2.8 mm. Body black, elytra and sometimes pronotum dark brown. Legs and antennae yellow, apical segments slightly darker. Upper surface thickly clothed with whitish, closely pressed hairs. Head, pronotum and elytra finely punctured, shining between the punctures. Abdominal tergites matt with dense, granular microsculpture. Aedeagus with long parameres, internal sclerites with complex folding patterns (see Fig. 112). This species is easily recognised by the wide, flat pronotum.
Habitat: Compost heaps.
Distribution: Scattered over England, N. Wales, S. Scotland, and E. Ireland. Possibly declining.
Biology: Recorded at light (Allen, 1985; Lane & Mann, 2006).

2. *Carpelimus lindrothi* Palm, 1942

Length 2-2.3 mm. Body black and shiny. Legs black with pale knees and tarsi. Antennae all black. Pubescence short. Abdominal tergites with cellular microsculpture. Aedeagus with long parameres, internal sclerites with complex folding patterns (see Fig. 112).
Similar species: Resembles *C. obesus* but smaller and with more transverse antennal segments.
Habitat: Large expanses of bare mud by reservoirs and rivers.
Distribution: A recent immigrant from mainland Europe, first found in Britain in 1976 (Owen, 1979). Now known from East Anglia, S. England and the Midlands.
Biology: Recorded at light (Lane & Mann, 2006).

3. *Carpelimus obesus* (Kiesenwetter, 1844) – Plate 27

Length 2.5-3.5 mm. Body black and shiny. Head and pronotum thickly clothed with whitish, closely pressed hairs. Elytra with more erect yellowish hairs. Head, pronotum and elytra densely punctured. Abdominal tergites with wrinkly microsculpture. Aedeagus with long parameres, internal sclerites with complex folding patterns (see Fig. 112).
Similar species: Resembles a dark *C. rivularis* or *C. bilineatus*, but more strongly punctured and without produced temples.
Habitat: Muddy margins of rivers and ponds, on sandy riverbanks in mainland Europe.
Distribution: A recent immigrant from mainland Europe, first recorded in Britain in 1948 (Steel, 1953). Now found over much of England except for the South-West. Also collected in Ireland in 1950 (Nash *et al.*, 1997).
Biology: Recorded at light (Lane & Mann, 2006).

4. *Carpelimus pusillus* (Gravenhorst, 1802)

Length 2-2.3 mm. Body black, elytra brown to dark brown. Legs mainly yellow. Colour of antennae variable. Pubescence whitish, close pressed on elytra. Head and pronotum finely and densely punctured, punctures only appearing distinct at high magnifications. Punctures on abdominal tergites slightly sparser towards apex, microsculpture lacking. Aedeagus with long parameres (see Fig. 112).
Similar species: This species is most likely to be confused with *C. corticinus* or *C. manchuricus*, but it is slightly smaller with paler elytra and has larger eyes.
Habitat: In the north on margins of rivers with a moderate flow. Also found regularly in litter piles away from rivers. Recorded in numbers on sewage-sludge drying beds (Green, 1983).
Distribution: England, Wales, S. Scotland, E. and SW. Ireland.
Biology: Recorded at light (Allen, 1985; Lane & Mann, 2006).

Subgenus *Myopinus* Scheerpeltz, 1937

5. *Carpelimus elongatulus* (Erichson, 1839) – Plate 28

Length 2-3 mm. Body black, elytra and pronotum sometimes dark brown. Legs red, occasionally with darker femora. Antennae with basal segments clear red, apical segments often darker. Diameter of eyes shorter than strongly expanded temples. Pubescence yellowish. Head and pronotum with strong, granular microsculpture except for small shiny area around the suture between the frons and clypeus. Elytra closely punctured with punctures directed backwards. Abdominal tergites with cellular microsculpture. Aedeagus relatively large for its body size with characteristic internal sclerites (see Fig. 112). This species is easily recognised by its large rounded temples, small eyes and matt head and pronotum.

Habitat: The most catholic *Carpelimus* species, found in a wide variety of wetland and riparian environments.

Distribution: Lowland areas of England, Wales, C. Scotland and E. Ireland.

Subgenus *Paratrogophloeus* Hatch, 1957

6. *Carpelimus bilineatus* Stephens, 1834 – Plate 29

Length 3-4 mm. External appearance similar to *C. erichsoni*.

Similar species: Only distinguishable from *C. erichsoni* on the aedeagus (see Fig. 112).

Habitat: Uncertain, but apparently similar to *C. erichsoni*. Also recorded from a heap of grass cuttings.

Distribution: So far recorded from England and Scotland. Irish records have yet to be checked since the recognition of *C. erichsoni*. Apparently scarcer than *C. erichsoni*.

7. *Carpelimus erichsoni* Sharp, 1871

Length 3-4 mm. Body black, elytra and pronotum rarely dark brown. Legs red, occasionally with slightly darker femora. Antennae with basal segments clear red, apical segments often darker. Head and pronotum thickly clothed with closely pressed hairs. Elytra with rather short, more erect yellowish hairs. Head, pronotum and elytra shining between the punctures. Abdominal tergites with cellular microsculpture. Aedeagus with short parameres, callipers and rosettes prominent among the internal sclerites (see Fig. 112).

Similar species: Previously confused with *C. bilineatus*. The size and shape of the pronotum make this species pair distinct within the genus. Small specimens may be confused with *C. rivularis* or *C. similis*, but they can be recognised by the wide marginal band of coarse puncturation on the pronotum and stronger puncturation on the elytra.

Habitat: Found in a wide variety of wetland and riparian environments with silty substrates.

Distribution: Mainly in lowland areas of England, Wales and Scotland, where it is one of the more frequently encountered *Carpelimus* species. Irish records of *C. bilineatus* could refer to this species.

Biology: Either this species or *C. bilineatus* frequently recorded at light (Allen, 1985; Lane & Mann, 2006).

8. *Carpelimus rivularis* (Motschulsky, 1860)

Length 2.5-3.2 mm. Body black to brown-black. Colour of legs and antennae variable. Head and pronotum thickly clothed with closely pressed hairs. Elytra with rather short, more erect yellowish hairs. Head, pronotum and elytra shining between the punctures. Abdominal tergites with cellular microsculpture. Aedeagus with short parameres, only the central hook prominent among the internal sclerites (see Fig. 112).

Similar species: See *C. erichsoni*, *C. similis* and *C. impressus*.

Habitat: Found in a wide variety of wetland and riparian environments with silty substrates.

Distribution: Widespread in lowland areas of England, Wales, S. Scotland and Ireland. Probably the most commonly encountered *Carpelimus* species.

Biology: Recorded at light (Lane & Mann, 2006).

9. *Carpelimus similis* Smetana, 1967

Length 3-4 mm. Body black, elytra and pronotum rarely dark brown. Legs red, occasionally with slightly darker femora. Antennae with basal segments clear red, apical segments often darker. Head and pronotum thickly clothed with closely pressed hairs. Elytra with rather short, more erect yellowish hairs. Head, pronotum and elytra shining between the punctures. Abdominal tergites with cellular microsculpture. Aedeagus with short parameres, callipers and central hook prominent among the internal sclerites (see Fig. 112).

Similar species: Previously confused with *C. bilineatus*, from which, apart from the key characters, it can be distinguished by the finer punctures on the elytra and the flatter pronotum with shallower depressions. Somewhat intermediate between *C. bilineatus* / *C. erichsoni* and *C. rivularis*.

Habitat: Found on a wide variety of mineral substrates in riparian environments.

Distribution: Scattered over S. England, Wales and the Midlands. Recently recorded from Ireland (Regan & Anderson, 2004).

Biology: Recorded at light (Allen, 1994) and in flight interception traps (Levey, 2007).

Subgenus *Troginus* Mulsant & Rey, 1878

10. *Carpelimus incongruus* Steel, 1969 – Plate 31

Length 2.1-2.8 mm. Body black, pronotum and elytra sometimes dark brown. Legs dark red. Antennae dark. Density of punctures somewhat variable, but upper surface generally shining between the punctures. Pronotum with very weak depressions, but with faint longitudinal ridge along centre line Front margin of labrum deeply incised, appearing bilobed. Aedeagus as in Fig. 112. Head wider in males. This species and *C. zealandicus* are together very distinctive within the genus by virtue of the form of the head and pronotum. They were previously confused under the name *C. zealandicus* (P.M. Hammond, pers. comm.) and investigations are continuing in order to establish their true identities and distinguishing characters. This species was originally recorded in Britain as a new species by Steel (1969). In view of the taxonomic uncertainties surrounding the *C. zealandicus* complex, Steel's name has been reinstated in this key pending their resolution.

Habitat: Sandy substrates in quarries, by rivers and in coastal environments.

Distribution: A recent introduction from New Zealand, first recorded in 1968 and now spread over England, Wales, S. Scotland and, most recently, N. Ireland (leg. R. Anderson).

11. *Carpelimus schneideri* (Ganglbauer, 1895)

Length 1.7-2.25 mm. Body black, elytra dark brown. Legs red, antennae darker. Eyes slightly longer than temples, which are rather flat and hardly differentiated from the thick neck. Pronotum expanded toward front, sides less rounded than in most other *Carpelimus* species, depressions absent. Pubescence white, on elytra short and closely pressed, almost scale-like. Microsculpture on head and pronotum strong and granular, becoming flat and cellular at back of head. Abdominal tergites lacking microsculpture between the fine punctures. Punctures on elytra and abdominal tergites directed backwards.

Habitat: On mud in estuaries and saltmarshes.

Distribution: Solway, Norfolk (leg. H.W. Daltry), Kent, Isle of Man. Rarely recorded.

12. *Carpelimus zealandicus* (Sharp, 1900)

Length 2.5-3 mm. Colour and general features similar to *C. incongruus*. Aedeagus as in Fig. 112. Previously confused with *C. incongruus* and only recently recognised in Britain (P.M. Hammond, pers. comm.).

Similar species: On average larger than *C. incongruus* with broader elytra and abdomen in comparison with head and pronotum. Pubescence on elytra longer.

Distribution: A recent introduction from New Zealand, established by 1997 (Yeates & Williams, 2006). Recorded from Surrey, Berkshire, Isles of Scilly, Yorkshire, S. Scotland (M. Telfer, R. Booth, R. Lyszkowski pers. comm.), but probably still expanding its range.

Subgenus *Trogophloeus* Mannerheim, 1830

13. *Carpelimus corticinus* (Gravenhorst, 1806) – Plate 30

Length 2-2.5 mm. Body black, elytra and pronotum very rarely brown. Legs dark, occasionally with red tibiae, very rarely completely yellow. Antennae with all segments dark, rarely clear red. Head and pronotum closely punctured, but with shiny intervals, slightly less closely punctured on clypeus. Abdominal tergites with cellular microsculpture. This species is variable in the form of its head and pronotum, but can be recognised on the smooth convexity of its temples, which are roughly the same size as the eyes, and the distinct punctures on the pronotum. Aedeagus with characteristic internal sclerites (see Fig. 112).

Similar species: See *C. manchuricus*.

Habitat: Found in a wide variety of wetland environments with silty substrates, but not normally associated with the main channels of rivers, unless their flow is very sluggish.

Distribution: Widespread in England, Wales and Ireland, but rarer in Scotland and absent from much of NE. Scotland. One of the more frequently encountered *Carpelimus* species.

Biology: Recorded at light (Lane & Mann, 2006).

14. *Carpelimus foveolatus* (Sahlberg, 1832)

Length 1.8-2 mm. Body black. Legs black with the tarsi, knees and the extreme apices of the tibiae paler. Pubescence short. Head and pronotum with strong, granular microsculpture except for small shiny area around the suture between the frons and clypeus. Elytra strongly punctured with shiny intervals between the punctures. Abdominal tergites with cellular microsculpture. Furrows on pronotum represented by four distinct depressions, of which the front pair are oblique. Very distinctive within the genus by virtue of its small size, dark colour and shining elytra, which contrast with the matt head and pronotum. Aedeagus with internal sclerites weakly chitinised (see Fig. 112).

Habitat: Coarse grass and tidal litter in upper saltmarsh.

Distribution: English coast from Somerset to Yorkshire.

15. *Carpelimus gracilis* (Mannerheim, 1830)

Length 1.8-2 mm. Body black, elytra brown to dark brown. Legs and antennae yellow to red, antennae with apical segments often darker. Pubescence whitish. Head and pronotum finely punctured, head more densely punctured except for the clypeus, where the punctures are much sparser. Abdominal tergites densely punctured towards base and more sparsely punctured towards apex. Depressions on pronotum rather shallow. Aedeagus with characteristic internal sclerites (see Fig. 112).

Similar species: Together with *C. subtilis*, this species is easily recognised by its small size and narrow form.

Habitat: Mostly on sand and shingle by rivers and streams, although it appears to be a good disperser and stray specimens are liable to turn up anywhere. Somewhat cryptic in its habits and possibly overlooked at many sites.

Distribution: Scattered over England, S. Wales, C. Scotland, and E. Ireland.

Biology: Recorded at light (Lane & Mann, 2006).

16. *Carpelimus halophilus* (Kiesenwetter, 1844)

Length 1.5-2 mm. Body black, elytra red to dark brown. Legs red, femora and tibiae sometimes darkened. Antennae red to black, basal segments occasionally lighter than apical segments. Pubescence rather short. Head and pronotum densely and finely punctured, the surface appearing granular except on clypeus, which is much more sparsely punctured. Elytra strongly and roughly punctured. Abdominal tergites without microsculpture, more sparsely punctured toward apex. Pronotum narrow, somewhat uneven, but lacking obvious depressions. Aedeagus with internal sclerites weakly chitinised (see Fig. 112). A rather variable species.

Habitat: Seashore, estuaries and saltmarsh.

Distribution: Coasts from S. Wales south and east to Norfolk.

17. *Carpelimus impressus* (Boisduval & Lacordaire, 1835)

Length 2.5-3 mm. Body black, elytra and pronotum rarely dark brown; colour of legs and antennae variable, but basal segment of antennae usually clear red. Head and pronotum thickly clothed with closely pressed hairs. Elytra with rather short, more erect yellowish hairs. Head and pronotum densely punctured. Abdominal tergites with cellular microsculpture. Aedeagus with characteristic internal sclerites similar to the rosettes found in members of the subgenus *Paratrogophloeus* (see Fig. 112).

Similar species: Somewhat intermediate in general appearance between *C. rivularis* and *C. corticinus* or *C. manchuricus*. Head and pronotum more closely punctured and duller than *C. rivularis*. Eyes larger and head more transverse than *C. corticinus* and *C. manchuricus*.

Habitat: Undisturbed, shaded floodplain wetlands and carr.

Distribution: S. and E. England north to Yorkshire. Widespread, but scattered in Ireland.

18. *Carpelimus manchuricus* (Bernhauer, 1938)

Length 2-2.5 mm. Body black, rarely with elytra dark brown. Legs usually darkened centrally on the femora and tibiae. Antennae dark, rarely with basal segment red. Punctures and microsculpture on head and pronotum somewhat variable, but rarely with shiny intervals. Abdominal tergites with cellular microsculpture. Aedeagus with long parameres, internal sclerites weakly chitinised (see Fig. 112). British populations belong to the subspecies *subtilicornis* (Roubal).

Similar species: Formerly confused with *C. corticinus*, this species is distinguished by its larger eyes and the surface sculpture on the pronotum.

Habitat: Exposed fine sand and silt by rivers. According to Hammond (1998b) associated with large lowland rivers and their tributaries.

Distribution: England, E. Wales and C. Scotland.

19. *Carpelimus subtilis* (Erichson, 1839)

Length 1.8-2 mm. Body dark brown with paler elytra. Legs and antennae yellow. Pubescence whitish. Puncturation on head and pronotum more or less uniform. Abdominal tergites densely punctured towards base and more sparsely punctured towards apex. Eyes shorter than temples. However, there are reports of large-eyed forms in mainland Europe. Depressions on pronotum rather shallow. The aedeagus shown in Fig. 112 is drawn from specimens from mainland Europe. In the few aedeagi examined from Britain, the internal sclerites were not well chitinised and difficult to see.

Similar species: This species is most likely to be confused with *C. gracilis*, from which it can normally be recognised by the smaller eyes, paler colour and the internal sclerites of the aedeagus.

Habitat: Adapted to life in interstitial habitats in shingle and sand. Usually by streams and rivers but odd specimens can turn up in a variety of locations.

Distribution: Scattered over England, Wales and S. Scotland. Rarely recorded.

Biology: Recorded from a flight interception trap (Levey, 2007).

10. *MANDA* Blackwelder, 1952

Fairly distinctive within the Oxytelinae, *Manda* is similar in form to *Planeustomus*, but is much larger. There are five tarsal segments. The genus contains three species and is restricted to the Holarctic region. Only one species has been recorded in Britain.

1. *Manda mandibularis* (Gyllenhal, 1827) – Plate 32

Length 5-6.5 mm. Body predominantly reddish brown, head black, pronotum often darkened centrally and elytra often darkened toward apex. Legs and antennae red. Pubescence sparse, but long and semi-erect, with longitudinal stripes on elytra orientated in alternating directions from one stripe to the next. Upper surface covered by granular microsculpture. Antennae thickened toward apex, segments 6 to 10 roughly quadrate. Mandibles produced forwards. Eyes more than 4x longer than the straight temples which form a distinct angle with neck (see Fig. 12). Pronotum with a rounded, unpunctured, longitudinal central ridge and a raised unpunctured prominence on either side. Each elytron with four rounded ridges (including the one along suture), which disappear toward apex, punctures irregular. Abdominal tergite VIII with concave hind margin.

Habitat: At edges of woodland ponds with fluctuating water levels.

Distribution: S. England. Recent records from Gloucestershire and Sussex only. Can be abundant when found.

11. *OCHTHEPHILUS* Mulsant & Rey

The long elytra in comparison to the pronotum are very characteristic. The pronotum has a central longitudinal ridge, two basal, obliquely transverse ridges on each side and normally a small boss on each side in the front half that is sometimes absent. There is a large variation in the height of these ridges within each species. The upper surface is pubescent, the hairs being relatively long on the abdominal tergites and short on the elytra. Cellular microsculpture is evident to varying degrees on the head, pronotum and abdominal tergites, but absent on the elytra. The terminal segment of the maxillary palps has a characteristic shape, being tapered due to a sinuate external margin (see Fig. 115). Abdominal tergite VIII has a concave hind margin in both sexes. There are no obvious external differences between the sexes. The male genitalia are not well sclerotised and are subject to distortion, if they become desiccated. They are best mounted in a medium such as DMHF.

This genus is almost exclusively riparian in its habitats and usually associated with fast to moderately fast-flowing streams and rivers on a variety of substrates.

Ochthephilus contains 48 species and is restricted to the Holarctic and Oriental regions. Four species are reliably recorded from Britain. Two of them are also recorded from Ireland. Makranczy (2001) revised the Central European species. The British species were reviewed by Lott (2008a), who also provided brief ecological notes.

Key to species of *Ochthephilus*

1. Antennal segments 6-10 quadrate to elongate; side margins of abdominal tergite VIII sinuate just in front of sharply pointed hind angles (Fig. 113); length > 4 mm; [genital tergite entire] ....
.... 3. *aureus* (Fauvel) (p. 75)

— Antennal segments 6-10 quadrate to transverse; side margins of abdominal tergite VIII simple in front of less sharply pointed hind angles (Fig. 114); length < 4 mm 2

2. Diameter of eyes less than 2x length of temples as seen from above (Fig. 115); antennal segment 6 as large as 5 and 7; aedeagus with short parameres (see Fig. 119); [pronotum matt through strong surface sculpture; genital tergite entire as in Fig. 118; elytra and appendages pale brown, rarely darker] ....
.... 2. *angustior* Bernhauer (p. 75)

— Diameter of eyes at least 2x length of temples as seen from above (Fig. 116); antennal segment 6 smaller than both 5 and 7; parameres longer (see Fig. 119); [pronotum more shiny though still with surface sculpture; elytra often dark brown, occasionally lighter] 3

3. Genital tergite deeply incised at apex, females with central projection in incision (Fig. 117); first antennal segment usually darkened .... 4. *omalinus* (Erichson) (p. 75)

— Genital tergite entire (Fig. 118); first antennal segment pale ....
.... 1. *andalusiacus* (Fagel) (p. 75)

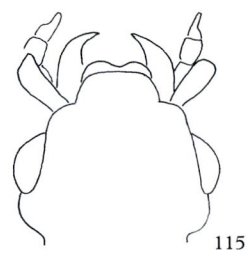

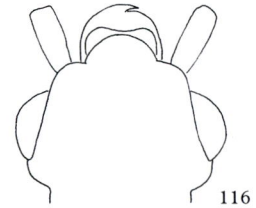

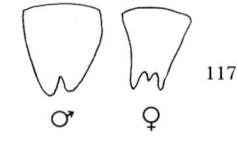

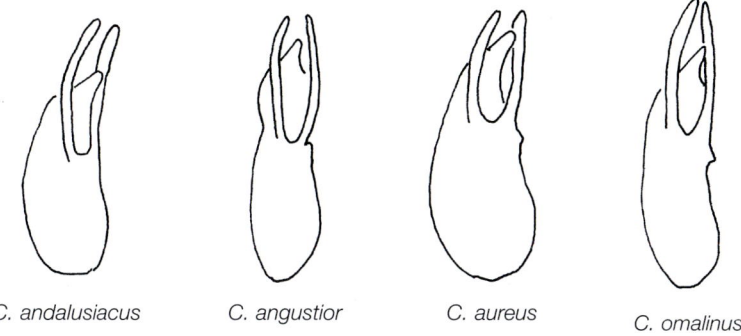

C. andalusiacus   C. angustior   C. aureus   C. omalinus

Fig. 119. The aedeagi of *Ochthephilus* species.

1. *Ochthephilus andalusiacus* (Fagel, 1957)

Length 3.5-4 mm. Body dark brown to black, elytra variable. Legs and antennae (at least at base) yellow. Aedeagus with long parameres (see Fig. 119).
Similar species: Resembles *O. omalinus*, but often has paler antennae.
Habitat: Stream margins, usually with fast flow, often with *O. aureus*.
Distribution: Recorded from SW. England, Midlands, N. England and C. Scotland.

2. *Ochthephilus angustior* Bernhauer, 1943

Length 3-3.5 mm. Body dark brown, usually with paler elytra. Legs and antennae (at least at base) yellow. Aedeagus with short parameres (see Fig. 119). Surface microsculpture on pronotum noticeably stronger than in other *Ochthephilus* species and body shorter and narrower, elytra often paler.
Habitat: Sandy margins of mature rivers, often with *O. omalinus*.
Distribution: W. Midlands, N. England and more rarely in Scotland.

3. *Ochthephilus aureus* (Fauvel, 1871) – Plate 33

Length 4-5 mm. Body dark brown to black, sometimes with paler elytra. Legs yellow. Antennal colour variable. Aedeagus with short parameres (see Fig. 119).
Similar species: Individuals with shorter antennae can be distinguished from *O. omalinus* by a combination of their large size and smaller eyes, whose diameter is less than the length of the temple when seen from above.
Habitat: Banks of streams and rivers including those in caves. Kasule (1966) reported finding a larva, possibly of this species, on wet sticks.
Distribution: Widespread in Britain and Ireland, but rare or absent in many lowland areas.

4. *Ochthephilus omalinus* (Erichson, 1840)

Length 3.5-4 mm. Body dark brown to black, occasionally with paler elytra. Legs and antennae darkened. Aedeagus with short parameres (see Fig. 119). The only British species with an incised genital tergite.
Similar species: Pronotum tends to be broader than *O. andalusiacus* and *O. angustior*, sometimes with a more even surface due to less pronounced ridges.
Habitat: Riverbanks.
Distribution: Possibly the most frequently encountered member of the genus. Widespread and sometimes abundant in Britain but absent from lowland areas of SE. and E. England. More rarely recorded in Ireland (N. and E. only).
Biology: Observed flying to breeding habitat in the spring (Lott, 2008a).

12. *PLANEUSTOMUS* Jacquelin du Val, 1857

Small body size coupled with body form and punctured striae on the front half of the elytra are enough to recognise this genus. Upper body surface covered with cellular microsculpture. Antennae expanded towards apex. Mandibles produced forwards. Two curved longitudinal rows of punctures present on the centre of the pronotum. Like *Manda*, the pronotum has a rounded, unpunctured, longitudinal central ridge and a raised unpunctured prominence on either side, but unlike *Manda*, there are no regular ridges on the elytra. Elytra with rows of hairs pointing diagonally inwards on inner half, backwards in the centre and diagonally outwards on outer half. Abdominal tergites VII and VIII with concave apical margins. This genus contains 24 described species and is distributed in the Palaearctic, Afrotropical and Oriental regions. Two species have been recorded from Britain.

Key to species of *Planeustomus*

1. Eyes larger; elytral striae more regular in front half of elytra; elytra nearly twice as long as pronotum (Fig. 120) .... 2. *palpalis* (Erichson) (p. 77)

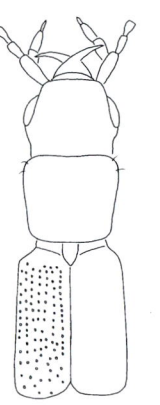

120

- Eyes smaller; elytral striae more irregular; elytra less than 1.5x longer than pronotum (Fig. 121) .... 1. *flavicollis* Fauvel (p. 76)

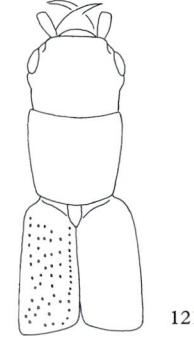

121

1. *Planeustomus flavicollis* Fauvel, 1871

Length 2-2.5 mm. Body yellow-brown, elytra yellow, apical abdominal tergites sometimes darkened. Legs and antennae yellow.

Similar species: According to some keys, this species can be distinguished by its paler colour, but *P. palpalis* can be variable in its colour and there is some overlap between the two species. However, the characters given in the key should separate the two species comfortably. In addition the pubescence on the elytra and the marginal setae on the pronotum are much shorter.

Habitat: Uncertain. Recorded mainly in the New Forest from flood refuse by a ditch (Sharp, 1912) and more recently in numbers from leaf litter (Giusti, 2007). Possibly subterranean and therefore difficult to find.

Distribution: Surrey and the New Forest. Extremely rare and currently known from just a handful of sites in the world.

2. *Planeustomus palpalis* (Erichson, 1839) – Plate 34

Length 2-2.5 mm. Head and apex of abdomen dark brown. Pronotum and base of abdomen pale brown. Elytra yellow. Legs yellow. Antennae yellow, often darker toward apex.
Habitat: In soft sediments by fresh water, often in shade. Steel (1949) reported it on two occasions in a garden and speculated that it may have been breeding under turf in a lawn that tended to hold water after rain.
Distribution: SE. England; sometimes abundant, but rarely encountered.
Biology: Disperses by flight during the evening (Twinn, 1958). Also reported from a flight interception trap (Welch, 1992).

13. *TEROPALPUS* Solier, 1849

Resembles a rather large, hairy *Carpelimus* but with reduced depressions on the pronotum. Maxillary palps with enlarged penultimate segment and needle-like terminal segment, like *Carpelimus*. This genus contains nine described species and is native to the Australian, Neotropical and Nearctic regions. One introduced species has been recorded in Britain and Ireland.

1. *Teropalpus unicolor* (Sharp, 1900) – Plate 35

Length 3-3.5 mm. Body black. Legs and antennae dark red with tarsi and extremities of tibiae clear red. Head, pronotum and elytra finely and closely punctured, very closely on pronotum, with dense white pubescence. Microsculpture not evident between the punctures. Antennae with most segments elongate. Eyes nearly 1.5x longer than gently rounded temples. Depressions on pronotum represented by four shallow impressions. Punctures on abdominal tergites directed backwards. Hind margin of abdominal tergite VIII slightly convex.
Habitat: Found under seaweed in drift-line on sandy beaches.
Distribution: Coastal from S. Wales east to Essex. E. Ireland. Introduced from New Zealand (Yeates & Williams, 2006) and established since the end of the 19th century (Keys, 1918).

14. *THINOBIUS* Kiesenwetter, 1844

The smallest species in the subfamily belong to this genus. The depressions on the pronotum are usually absent or at most barely suggested. The penultimate segment of the maxillary palps is expanded and the terminal segment minute, so that it is barely visible. The hind margin of the abdominal tergite VIII is concave but generally straightened in the centre. There are no obvious external differences between the sexes.

Thinobius species are adapted for life in the gaps between the particles of fine shingle and sand by water margins. With the exception of one species, they are only recorded from bare, exposed sediments in fast-flowing streams and rivers or, more infrequently, lake margins. Their small size distinguishes them from most other genera in this habitat. *Carpelimus gracilis*, *C. subtilis*, *Hydrosmecta delicatula*, *H. subtilissima* and various *Meotica* species are other small Staphylinidae that sometimes share their habitat in Britain. Because of their subterranean habitat and small size, their capture often requires special techniques, such as the submersion of gravel in water.

The British species of *Thinobius* fall into two morphological groups. The *T. longipennis* group (*T. brevipennis, T. ciliatus, T. crinifer, T. longipennis*) are fairly distinctive by virtue of their stout body form and characteristic head shape. When floating on the surface of the water they attempt to stand clear of the surface film and are apt to take flight in warm weather. Their bodies are less flexible than other *Thinobius* species, which, on contact with the water, often lie on their sides curled up in an S rather like species of *Hydrosmecta*. Species of the *T. longipennis* group tend to live in colonies like *Bledius* species. They can occur in large numbers, but are frequently localised within the shingle bank.

Worldwide there are 124 described species of *Thinobius*, of which seven are currently recorded from Britain. Two species have been recorded from Ireland, but the true identity of one of them awaits clarification. Lott (1993) reviewed the British species allied to *T. longipennis* (*T. crinifer, T. longipennis* and *T. praetor*) following Smetana (1959). The central European species of the *Thinobius linearis* group (of which *T. bicolor* is a member) were revised by Makranczy & Schülke (2001). *T. bicolor* was not included in this revision, but is conspecific with the species on mainland Europe recognised as *T. linderianus* Scheerpeltz (G. Makranczy, pers. comm.). The name, *T. bicolor*, has priority.

Key to species of *Thinobius*

1. Head rectangular; temples parallel or expanded behind eyes and longer than diameter of eyes when seen from above (see Figs 122-124); head, pronotum and elytra with distinct punctures, often with shining intervals between the punctures; body more slender and parallel-sided 2

- Head more oval due to smoothly rounded temples, which are shorter than diameter of eyes when seen from above (see Figs 125, 126); head, pronotum and elytra matt without distinct punctures; body more stocky 4

2. Head, pronotum and elytra unicolorous yellow; temples more than 1.5x as long as eyes when seen from above; head as long as pronotum (Fig. 122); antennal segments 5 to 6 transverse; punctures on head and pronotum finer with intervals less shiny .... 7. *newberyi* Scheerpeltz (p. 81)

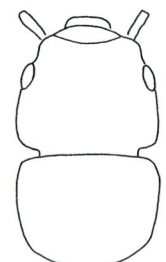

122

- Head and pronotum brown, darker than elytra; temples equal to or less than 1.5x as long as eyes when seen from above; head longer than pronotum (see Figs 123, 124); antennal segments 5 and 6 quadrate or elongate; punctures on head stronger with shiny intervals 3

3. Antennal segments 5 to 8 quadrate to transverse; temples about 1.5x as long as diameter of eyes when seen from above; elytra shorter than head and pronotum combined (Fig. 123) .... 1. *bicolor* Joy (p. 80)

123

- Antennal segments 5 to 8 elongate; diameter of eyes almost as long as temples when seen from above; elytra longer than head and pronotum combined (Fig. 124) .... 6. *major* Kraatz (p. 81)

124

4. Pronotum smaller in comparison to head (Fig. 125); length smaller, *c* 1 mm .... 2. *brevipennis* Kiesenwetter (p. 80)

125

- Pronotum larger in comparison to head (Fig. 126); length greater, *c* 1.5 mm (three species most easily separated using the male genitalia) 5

126

5. Second segment of antennae nearly twice as long as third segment; apical process of aedeagus longer than basal part and strongly sinuate in profile (Fig. 127) .... 3. *ciliatus* Kiesenwetter (p. 81)

- Second segment of antennae less than 1.5x longer than third segment; apical process of aedeagus absent or much shorter than basal part and only slightly curved in profile 6

6. Aedeagus with apical process (Fig. 128) ....
.... 4. *crinifer* Smetana (p. 81)

- Aedeagus without apical process (Fig. 129) ....
.... 5. *longipennis* (Heer) (p. 81)

1. *Thinobius bicolor* Joy, 1911 – Plate 37

Length 1.8 mm. Head, pronotum and abdomen brown. Legs, antennae and elytra yellow. Upper surface with short, closely pressed pubescence. Elytra with scattered, short, erect setae in addition. Head, pronotum and elytra almost uniformly punctured, shiny between the punctures with no microsculpture visible at 80x magnification. Abdominal tergites with punctures directed backwards and weak cellular microsculpture. Head with two longitudinal furrows on frons, temples weakly expanded behind eyes, almost parallel-sided.
Similar species: Similar in size and general shape to small *Hydrosmecta* species (Aleocharinae) and could be overlooked in the field, but easily distinguished by subfamily and generic characters on close examination. See also *T. major* and *T. newberyi*.
Habitat: Sand and shingle by streams and rivers.
Distribution: Locally distributed in SW. England, S. Wales, N. England, Scotland and N. Ireland. Unlike the *T. longipennis* group it is rarely recorded in large numbers.

2. *Thinobius brevipennis* Kiesenwetter, 1850

Length 1 mm. Body black. Legs and antennae dark.
Similar species: Similar to species in the *T. longipennis* group, but much smaller with antennal segments 4 to 10 more transverse and eyes smaller in relation to the temples.
Habitat: Occurs on much finer sediments than its congeners; associated with both vegetated and unvegetated wet sand and silt.
Distribution: S. England, East Anglia, N. Wales, mostly near the coast. Also an old record from Lancashire. Rarely recorded.

3. *Thinobius ciliatus* **Kiesenwetter, 1844** – Plate 38

Length 1.5 mm. Head, pronotum and abdomen black. Elytra brown. Legs and antennae brown. Pubescence and microsculpture as *T. crinifer*.
Similar species: Males are best separated from *T. crinifer* and *T. longipennis* by the highly characteristic aedeagus. Also, in general, *T. ciliatus* has shorter antennae and the pronotum is less transverse.
Habitat: Sand and fine shingle by streams and rivers.
Distribution: Locally distributed in SW. England, S. Wales, N. England and Scotland.

4. *Thinobius crinifer* **Smetana, 1959**

Length 1.5 mm. Head, pronotum and abdomen black. Elytra brown. Legs and antennae brown. Upper surface matt due to very short, dense pubescence resembling the dusting found in many Diptera species. Neck shining. Abdominal tergites with cellular microsculpture and fringes of long hairs on the apical margins.
Habitat: Unvegetated fine shingle by streams and rivers.
Distribution: Locally distributed in S. Wales, N. England and Scotland.

5. *Thinobius longipennis* **(Heer, 1841)**

Length 1.5 mm. Head, pronotum and abdomen black. Elytra, legs and antennae brown. Pubescence and microsculpture as *T. crinifer*.
Similar species: Only distinguished with certainty from *T. crinifer* by reference to the male genitalia.
Habitat: Unvegetated, fine riverside shingle.
Distribution: Known only from the Coquet and Till catchments in Northumberland. Old Irish records of this species are likely to represent other more recently recognised species.

6. *Thinobius major* **Kraatz, 1857**

Length 2.25 mm. Head, pronotum and abdomen dark brown. Elytra yellow-brown, darker at base. Legs and antennae brown. Head and pronotum rather strongly and densely punctured. Elytra more finely punctured. Cellular microsculpture on abdominal tergites composed of large, transverse meshes. Furrows on head much less pronounced than in *T. bicolor*. Temples weakly rounded behind eyes, almost parallel-sided.
Similar species: Larger than *T. bicolor* and with longer elytra that are clothed with longer pubescence. Head and pronotum more strongly punctured.
Habitat: Sand and shingle by streams, rivers and lakes.
Distribution: Scottish Highlands. Rarely recorded.

7. *Thinobius newberyi* **Scheerpeltz, 1925**

Length 2 mm. Insect all yellow except for the darkened apical abdominal tergites. Upper surface finely and densely punctured, not very shiny. Pubescence on elytra denser than *T. bicolor*. Cellular microsculpture on abdominal tergites with transverse meshes. Head flat and trapezoidal with temples expanded behind eyes. Mid and hind tibiae angled on external margin.
Similar species: Probably closest to *T. bicolor,* but easily distinguished by colour, eye length, head shape and other characters mentioned above.
Habitat: Sand and shingle by running water.
Distribution: Wales, Cumbria and Scottish Highlands. Rarely encountered. Not yet recorded from outside Britain.

15. *THINODROMUS* Kraatz, 1857

Formerly included with *Carpelimus* under the name *Trogophloeus*, *Thinodromus* looks like a large *Carpelimus*, but the pronounced, unbroken, curved furrow behind two ovoid depressions on the pronotum is diagnostic. Maxillary palps with needle-like terminal segment, like *Carpelimus*. There are 98 described species worldwide, but only one has been recorded in Britain and Ireland. The Palaearctic species have been revised by Gildenkov (2001b).

1. *Thinodromus arcuatus* (Stephens, 1834) – Plate 36

Length 3-3.5 mm. Body black, closely punctured but shiny without microsculpture. Legs and antennae red-brown. The yellow pubescence is especially noticeable on the abdomen. Antennae with most segments elongate. Abdominal tergite VIII with concave hind margin.
Habitat: Associated with river and stream banks both on exposed mineral sediments and in moss on boulders and the walls of weirs.
Distribution: England, Wales, S. Scotland and Ireland. Rare or absent in some eastern areas.

References

Alexander, K.N.A. 2002. The invertebrates of living and decaying timber in Britain and Ireland – a provisional annotated checklist. *English Nature Research Reports* **467**.

Allen, A.A. 1949. *Oxytelus piceus* L. (Col., Staphylinidae) in the Isle of Wight and Kent. *Entomologist's monthly Magazine* **85**: 37.

Allen, A.A. 1974. *Bledius bicornis* Germ. (Col., Staphylinidae) in suburban West Kent. *Entomologist's monthly Magazine* **104**: 325.

Allen, A.A. 1985. *Carpelimus fuliginosus* (Grav.) (Col., Staphylinidae) at light in S.E. London. *Entomologist's monthly Magazine* **121**: 108.

Allen, A.A. 1994. *Carpelimus similis* (Smet.) (Col.: Staphylinidae) in S.E. London. *Entomologist's Record & Journal of Variation* **106**: 115-116.

Anderson, R. 1986. *Deleaster dichrous* (Grav.) (Col., Staphylinidae) associated with silo pits. *Entomologist's monthly Magazine.* **122**: 184.

Assing, V. & Schülke, M. 2007. Supplemente zur mitteleuropäischen Staphylinidenfauna (Coleoptera, Staphylinidae). III. *Entomologische Blätter für Biologie und Systematik der Käfer* **102**: 1-78.

Barbut, J. 1781. *Les genres des insectes de Linne: constates par divers echantillons d'insectes d'Angleterre, copies d'apres nature.* Londres [London]: J. Dixwell.

Beaver, R.A. 1967. Notes on the fauna associated with elm bark beetles in Wytham Wood, Berks – I. Coleoptera. *Entomologist's monthly Magazine* **102**: 163-170.

Betz, O. 1999. A behavioural inventory of adult *Stenus* species (Coleoptera: Staphylinidae). *Journal of Natural History* **33**: 1691-1712.

Betz, O., Thayer, M.K. & Newton, A.F. 2003. Comparative morphology and evolutionary pathways of the mouthparts in spore-feeding Staphylinoidea (Coleoptera). *Acta Zoologica* **84**: 179-238.

Bro Larsen, E. 1936. *Biologische Studien über die tunnelgrabenden Käfer auf Skallingen.* Copenhagen: Kommission Hos C. A. Reitzels.

Bro Larsen, E. 1952. On subsocial beetles from the salt-marsh, their care of progeny and adaptation to salt and tide. *Transactions of the 9th International Congress of Entomology, Amsterdam* **1**: 502-506.

Chuter, K. 2000. Some scarce beetles in Kent. *Coleopterist* **9**: 42-43.

Cooter, J. & Barclay, M.V.L. (eds) 2006. *A Coleopterist's Handbook.* London: Amateur Entomologist's Society.

Crowson, R.A. 1982. Observations on *Phyllodrepoidea crenata* (Gravenhorst) (Col., Staphylinidae). *Entomologist's monthly Magazine* **118**: 125-126.

Dettner, K. 1983. Isopropylesters as wetting agents from the defensive secretion of the rove beetle *Coprophilus striatulus* F. (Coleoptera, Staphylinidae). *Insect Biochemistry* **14**: 383-390.

Dettner, K. 1993. Defensive secretions and exocrine glands in free-living staphylinid beetles – their bearing on phylogeny (Coleoptera: Staphylinidae). *Biochemical Systematics and Ecology* **21**:143.

Dettner, K. & Schwinger, G. 1982. Defensive secretions of three oxytelinae rove beetles (Coleoptera: Staphylinidae). *Journal of Chemical Ecology* **8**: 1411-1420.

Dettner, K., Schwinger, G. & Wunderle, P. 1985. Sticky secretion from two pairs of defensive glands of rove beetle *Deleaster dichrous* (Grav.) (Coleoptera: Staphylinidae). *Journal of Chemical Ecology* **11**: 859-883.

Gildenkov, M.Yu. 2001a. *The Palaearctic Carpelimus Fauna (Coleoptera: Staphylinidae). The Problems of Species and the Formation of Species.* (2 vols.) Smolensk: Smolensk State Pedagogical University.

Gildenkov, M.Yu. 2001b. *Phylogenetic relations in the Oxytelinae subfamily. The Palaearctic Thinodromus Fauna (Coleoptera: Staphylinidae: Oxytelinae).* Smolensk: Smolensk State Pedagogical University.

Gildenkov, M.Yu. 2003. A new system of subfamily Oxytelinae (Coleoptera: Staphylinidae). *Kharkov Entomological Society Gazette* **10**: 32-38.

Giusti, A. 2007. Records of *Planeustomus flavicollis* Fauvel (Staphylinidae) and *Eutheia formicetorum* Reitter (Scydmaenidae) from the New Forest. *Coleopterist* **16**: 134.

Good, J.A. & Giller, P.S. 1991. The diet of predatory staphylinid beetles – a review of records. *Entomologist's monthly Magazine* **127**: 77-89.

Grebennikov, V.V. & Newton, A.F. 2009. Goodbye Scydmaenidae, or why the ant-like stone beetles should become megadiverse Staphylinidae *sensu latissimo* (Coleoptera). *European Journal of Entomology* **106**: 275-301.

Green, M.B. 1983. The macrofauna of sludge-drying beds. *Used-water Treatment* **2**: 261-299.

Hammond, P.M. 1968. *Oxytelus mutator* Lohse (Col., Staphylinidae) new to Britain. *Entomologist* **101**: 250-252.

Hammond, P.M. 1971. Notes on British Staphylinidae 2. – on the British species of *Platystethus* Mannerheim with one species new to Britain. *Entomologist's monthly Magazine* **106**: 93-111.

Hammond, P.M. 1998a. *Oxytelus migrator* Fauvel (Col., Staphylinidae), an Asian rove-beetle established in Britain. *Entomologist's monthly Magazine* **134**: 273-276.

Hammond, P.M. 1998b. Riparian and floodplain arthropod assemblages: their characteristics and rapid assessment. In: Bailey, R.G., José, P.V. &. Sherwood, B.R. (eds.) *United Kingdom Floodplains.* pp 237-282. Otley: Westbury.

Hammond, P.M. 2000. Coastal Staphylinidae (rove beetles) in the British Isles, with special reference to saltmarshes. In: Sherwood, B.R., Gardiner, B.G. & Harris, T. (eds.) *British Saltmarshes* pp. 247-302. Cardigan: Forrest Press.

Hansen, M. 1996. Catalogue of the Coleoptera of Denmark. *Entomologiske Meddelelser* **64**: 1-231.

Hansen, M. 1997. Phylogeny and classification of the staphyliniform beetle families (Coleoptera). *Royal Danish Academy of Science and Letters, Biologiske Skrifter* **48**: 1-339.

Herman, L.H. 1986. Revision of *Bledius* Part IV. Classification of species groups, phylogeny, natural history, and catalogue (Coleoptera, Staphyliniae, Oxytelinae). *Bulletin of the Ameican Museum of Naural History* **184**(1): 1-367.

Herman, L.H. 2001. Catalog of the Staphylinidae (Insecta, Coleoptera): 1758 to the End of the Second Millennium. *Bulletin of the American Museum of Naural History*, **265**: 1-4218.

Hinton, H.E. 1944. Some general remarks on sub-social beetles, with notes on the biology of the staphylinid, *Platystethus arenarius* (Fourcroy). *Proceedings of the Royal Entomological Society of London (A)* **19**: 115-128.

Hodge, P.J. 1979. *Bledius diota* Schiod. and *B. furcatus* (Ol.) in East Sussex. *Entomologist's monthly Magazine* **114**: 114.

Hyman, P.S. (revised Parsons, M.S.) 1994. *A review of the scarce and threatened Coleoptera of Great Britain.* Part 2. UK Nature Conservation: 12. Peterborough: Joint Nature Conservation Committee.

Kasule, F.K. 1966. The subfamilies of the larvae of Staphylinidae (Coleoptera) with keys to the larvae of British genera of Steninae and Proteininae. *Transactions of the Royal Entomological Society of London* **118**: 261-283.

Kasule, F.K. 1968. Field studies on the life-histories of some British Staphylinidae (Coleoptera). *Transactions of the Society for British Entomology* **18**: 49-80.

Keys, J.H. 1918. A list of the maritime, sub-maritime and coast-frequenting Coleoptera of South Devon and South Cornwall with especial reference to the Plymouth district. *Journal of the Marine Biological Association of the United Kingdom* **11**: 497-513.

Krogerus, R. 1925. Lebensweise und Entwicklung einiger *Bledius*-Arten. *Acta Societatis pro Fauna et Flora Fennica* **56**: 3-25.

Lane, S.A. & Mann, D.J. 2006. Notes on Coleoptera recorded at a mercury vapour light trap in Warwickshire, August 2004. *Coleopterist* **15**: 79-91.

Lawrence, J.F. & Newton, A.F. 1982. Evolution and classification of beetles. *Annual Review of Ecology and Systematics* **13**: 261-290.

Lawrence, J.F. & Newton, A.F. 1995. Families and subfamilies of Coleoptera (with selected genera, notes, references and data of family group-names). In: Pakaluk, J. & Slipinski, S.A. (eds.) *Biology, Phylogeny and Classification of Coleoptera.* Museum of the Zoological Institute PAN, Warsaw: pp 779-1006.

Levey, B. 2007. Recent interesting Coleoptera records from Wales. *Coleopterist* **16**: 138-140.

Lipkow, E. & Betz, O. 2005. Staphylinidae and fungi. *Faunistische- Ökolische Mitteilungen* **8**: 383-411.

Löbl, I. 1997. *Catalogue of the Scaphidiinae (Coleoptera:Staphylinidae).* Geneva MHN. 190 pp.

Löbl, I. & Smetana, A. (eds.) 2004. *Catalogue of Palaearctic Coleoptera Volume 2 Hydrophilidae – Histeroidea – Staphylinoidea.* Stenstrup: Apollo Books. 942 pp.

Lohse, G.A. 1982. 13. Nachtrag zum Verzeichnis der mitteleuropäischen Käfer. *Entomologische Blätter für Biologie und Systematik der Käfer* **78**: 115-126.

Lott, D.A. 1993. The British species of the *Thinobius longipennis* (Heer) group (Coleoptera: Staphylinidae). *Entomologist's Gazette* **44**: 285-287.

Lott, D.A. 2003. An annotated list of wetland ground beetles (Carabidae) and rove beetles (Staphylinidae) found in the British Isles including a literature review of their ecology. *English Nature Research Reports* **488**.

Lott, D.A. 2008a. The British species of *Ochthephilus* Mulsant & Rey (Col., Staphylinidae). *Coleopterist* **17**: 17-22.

Lott, D.A. 2008b. The aedeagi of the British and Irish species of *Bledius* Leach (Col., Staphylinidae) occurring on riverbanks. *Coleopterist* **17**: 161-189.

Makranczy, G. 2001. Zur Kenntnis der mitteleuropäischen Arten der Gattung *Ochthephilus* Mulsant & Rey, 1856 (Coleoptera, Staphylinidae, Oxytelinae). *Entomologische Blätter für Biologie und Systematik der Käfer* **97**: 177-184.

Makranczy, G. & Schülke, M. 2001. Typenstudien an den mitteleuropäischen Vertreten der Artengruppe des *Thinobius linearis* Kraatz, 1857 (Coleoptera, Staphylinidae, Oxytelinae). *Entomologische Blätter für Biologie und Systematik der Käfer* **97**: 185-193.

Nash, R., Anderson, R. & O'Connor, J.P. 1997. Recent additions to the list of Irish Coleoptera. *Irish Naturalist's Journal* **25**: 319-325.

Newton A.F. 1984. Mycophagy in Staphylinoidea (Coleoptera). In: Wheeler, Q. & Blackwell, M. (eds.) *Fungus-Insect Relationships: Perspectives in Ecology and Evolution.* Columbia University Press.

Newton, A.F. & Thayer, M.K. 1992. Current classification and family-group names in Staphyliniformia (Coleoptera). *Fieldiana Zoology* **67**: 1-92.

Newton, A.F. & Thayer, M.K. 1995. Protopselaphinae new subfamily for *Protopselaphus* new genus from Malaysia, with a phylogenetic analysis and review of of the Omaliinae group of Staphylinidae including Pselaphidae (Coleoptera). In: Pakaluk, J. & Slipinski, S.A. (eds.) *Biology, Phylogeny and Classification of Coleoptera.* Museum of the Zoological Institute PAN, Warsaw: pp 219-320.

Omer-Cooper, J. & Tottenham, C.E. 1934. Coleoptera taken in the air at Wicken Fen. *Entomologist's monthly Magazine* **70**: 231-234.

Owen, J.A. 1979. *Carpelimus lindrothi* Palm new to Britain. *Entomologist's monthly Magazine* **114**: 102.

Owen, J.A. 1997. Some uncommon beetles from Headley Warren, Surrey. *Entomologist's Record & Journal of Variation* **109**: 301-307.

Philp, E. 1990. *Scaphisoma assimile* Erichson (Col: Scaphidiidae) in Kent. *Entomologist's Record & Journal of Variation* **102**: 116.

Pope, R.D. 1977. Kloet & Hincks. A Check List of British Insects. Part 3: Coleoptera and Strepsiptera. Second revised edition. *Handbooks for the Identification of British Insects* **11**(3), pp. xiv+105.

Regan, E. & Anderson, R. 2004. Terrestrial Coleoptera recorded in Ireland, May 2003. *Bulletin of the Irish Biogeographical Society* **28**: 85-132.

Sharp, D. 1912. Discovery of *Planeustomus flavicollis*, Fauv. in England. *Entomologist's monthly Magazine* **48**: 162-163.

Shirt, D.B. (ed.) 1987. *British Red Data Books: 2. Insects*. Peterborough: Nature Conservancy Council.

Smetana, A. 1959. Neue Arten der Gattung *Thinobius* Kiesw. aus Europa (Col., Staphylinidae). *Casopis Ceskoslovenské Spolecnosti Entomologické* **56**: 265-275.

Steel, W.O. 1949. *Planeustomus palpalis* (Erichs.) and *Scopaeus sulcicollis* (Steph.) (Col., Staphylinidae) in a garden in Middlesex. *Entomologist's monthly Magazine* **85**: 50.

Steel, W.O. 1953. A species of *Carpelimus* (Col., Staphylinidae) new to Britain. *Entomologist's monthly Magazine* **89**: 214.

Steel, W.O. 1969. A British *Carpelimus* (Col., Staphylinidae) new to science. *Entomologist's monthly Magazine* **105**: 70-72.

Steidle, J.L.M. & Dettner, K. 1995a. Abdominal gland secretion of *Bledius* rove beetles as an effective defence against predators. *Entomologia Experimentalis et Applicata* **76**: 211-216.

Steidle, J.L.M. & Dettner, K. 1995b. The chemistry of the abdominal gland secretion of six species of the rove beetle genus *Bledius*. *Biochemical Systematics and Ecology* **23**: 757-765.

Twinn, D.C. 1958. The evening flight period of Coleoptera. *Entomologist's monthly Magazine* **94**: 233.

Vorst, O. 2003. Nieuws over Nederlandse kortshildkevers 2 – Omaliinae, Oxytelinae (Coleoptera: Staphylinidae). *Entomologische Berichte* **63**: 147-156.

Welch, R.C. 1977. Coleoptera from Rothamsted light traps at Monks Wood National Nature Reserve, Cambridgeshire during 1976. *Entomologist's. Record & Journal of Variation* **89**: 195-198.

Welch, R.C. 1992. *Planeustomus palpalis* (Er.) from two Northamptonshire localities. *Entomologist's monthly Magazine* **128**: 192.

Williams, S.A. 2009. *Scaphium immaculatum* (Olivier) (Staphylinidae) rediscovered in Britain. *Coleopterist* **18**: 148.

Wright, S. 1990. *Bledius germanicus* Wagner new to Nottinghamshire. *Entomologist's. Record* **102**: 226.

Wyatt, T.D. 1986. How a subsocial intertidal beetle, *Bledius spectabilis*, prevents flooding and anoxia in its burrow. *Behavioral Ecology and Sociobiology* **19**: 323-331.

Wyatt, T.D. 1989. Parental care in the subsocial intertidal beetle, *Bledius spectabilis*, in relation to parasitism by the ichneumon wasp, *Barycnemis blediator*. *Behaviour* **110**: 76-92.

Wyatt, T.D. & Foster, W.A. 1988. Distribution of the intertidal saltmarsh beetle, *Bledius spectabilis*. *Ecological entomology* **13**: 453-464.

Wyatt, T.D. & Foster, W.A. 1989. Leaving home: predation and the dispersal of larvae from the maternal burrow of *Bledius spectabilis*, a subsocial intertidal beetle. *Animal Behaviour* **38**: 778-785.

Yeates, G.W. & Williams, P.A. 2006. Export of plant and animal species from an insular biota. *Ecological Studies* **186**: 85-100.

Index

Main entries and start of main sections are shown in **bold**. Synonyms are given in *italics*.

Colour plates

Plate 1.
Scaphidium quadrimaculatum

Plate 2.
Scaphium immaculatum

Plate 3.
Scaphasoma agaricinum

All scale bars = 1mm

Plate 4.
Saigonium quadricorne

Plate 5.
Coprophilus striatulus

Plate 6.
Deleaster dichrous

Plate 7.
Syntonium aeneum

Plate 8.
Anotylus rugosus

Plate 9.
Anotylus sculpturatus

All scale bars = 1mm

Plate 10.
Oxytelus fulvipes

Plate 11.
Oxytelus laqueatus

Plate 12.
Oxytelus sculptus

Plate 13.
Platystethus cornutus

Plate 14.
Platystethus arenarius

Plate 15.
Aploderus caelatus

All scale bars = 1mm

Plate 16.
Bledius spectabilis

Plate 17.
Bledius spectabilis (profile)

Plate 18.
Bledius spectabilis

Plate 19.
Bledius furcatus

Plate 20.
Bledius furcatus (profile)

Plate 21.
Bledius bicornis

All scale bars = 1mm

Plate 22.
Bledius gallicus

Plate 23.
Bledius opacus

Plate 24.
Bledius pallipes

Plate 25.
Bledius subterraneus

Plate 26.
Bledius fergussoni

All scale bars = 1mm

Plate 27.
Carpelmius obesus

Plate 28.
Carpelmius elongatulus

Plate 29.
Carpelmius bilineatus

Plate 30.
Carpelmius corticinus

Plate 31.
Carpelmius incongruus

Plate 32.
Manda mandibularis

All scale bars = 1mm

Plate 33.
Ochthephilus aureus

Plate 34.
Planeustomus palpalis

Plate 35.
Teropalpus unicolor

Plate 36.
Thinodromus arcuatus

Plate 37.
Thinobius bicolor

Plate 38.
Thinobius ciliatus

All scale bars = 1mm